Principles of Seismology

Principles of Seismology

Elijah Walker

R CALLISTO
REFERENCE

www.callistoreference.com

Callisto Reference,
118-35 Queens Blvd., Suite 400,
Forest Hills, NY 11375, USA

Visit us on the World Wide Web at:
www.callistoreference.com

ISBN: 978-1-64116-569-3 (Hardback)

Cataloging-in-Publication Data

Principles of seismology / Elijah Walker.
    p. cm.
Includes bibliographical references and index.
ISBN 978-1-64116-569-3
1. Seismology. 2. Geophysics. 3. Earthquakes. I. Walker, Elijah.
QE534.3 .P75 2022
551.22--dc23

TABLE OF CONTENTS

PREFACE

The purpose of this book is to help students understand the fundamental concepts of this discipline. It is designed to motivate students to learn and prosper. I am grateful for the support of my colleagues. I would also like to acknowledge the encouragement of my family.

The scientific study of earthquakes and the propagation of elastic waves through the Earth or other planet like bodies is referred to as seismology. It also includes the environmental effects of the earthquake such as tsunamis. The elastic waves that propagate in fluid or solid materials are known as seismic waves. Some of the different types of seismic waves studied within this field are body waves, surface waves and normal modes. Body waves travel through the interior of the materials. Surface waves travel along surfaces or interfaces between materials. Normal modes are a form of standing wave. This book provides significant information of this discipline to help develop a good understanding of seismology and related fields. It brings forth some of the most innovative concepts and elucidates the unexplored aspects of this field. Experts and students actively engaged in this field will find this book full of crucial and unexplored concepts.

A foreword for all the chapters is provided below:

Chapter – What is Seismology?

Seismology refers to the study of earthquakes and seismic waves. Theory of plate tectonics, seismic imaging and inversion fall under its domain. This chapter closely examines about seismology and its related aspects to provide an extensive understanding of the subject.

Chapter – Seismic Waves

Seismic waves are the resulting waves from earthquakes, volcanic eruptions, magma movement, large landslides and large man-made explosions. They can be categorized into surface waves, P-waves, S-waves, Stoneley waves, etc. The topics elaborated in this chapter will help in gaining a better perspective about these seismic waves.

Chapter – Seismic Methods and Measurements

Seismic methods include reflection seismology, seismic refraction, isoseismal map, passive seismic, plus minus method, etc. Seismometer, seismogram, geophone, accelerograph, etc. are some of the seismic instruments. This chapter has been carefully written to provide an easy understanding of these seismic methods and instruments.

Chapter – Earthquakes: An Integrated Study

Earthquake is the shaking of the Earth's surface due to the passage of seismic waves through Earth's rocks. Slow earthquake, blind thrust earthquake, megathrust earthquake, submarine earthquake, doublet earthquake, etc. are a few of its types. This chapter discusses these types of earthquakes and related aspects in detail.

Chapter – Subfields of Seismology

Engineering seismology, earthquake forecasting, archaeoseismology, paleoseismology, fault mechanics and seismic tomography are some of the fields studied under seismology. This chapter delves into these fields of seismology to provide an in-depth understanding of the subject.

Elijah Walker

CHAPTER 1

What is Seismology?

Seismology refers to the study of earthquakes and seismic waves. Theory of plate tectonics, seismic imaging and inversion fall under its domain. This chapter closely examines about seismol-ogy and its related aspects to provide an extensive understanding of the subject.

Seismology is the study of vibrations within Earth. These vibrations are caused by various events, including earthquakes, extra-terrestrial impacts, explosions, storm waves hitting the shore, and tidal effects. Of course, seismic techniques have been most widely applied to the detection and study of earthquakes, but there are many other applications, and arguably seismic waves provide the most important information that we have concerning Earth's interior. Before going any deeper into Earth, however, we need to take a look at the properties of seismic waves. The types of waves that are useful for understanding Earth's interior are called body waves, meaning that, unlike the surface waves on the ocean, they are transmitted through Earth materials.

Imagine hitting a large block of strong rock (e.g. granite) with a heavy sledgehammer. At the point where the hammer strikes it, a small part of the rock will be compressed by a fraction of a millime-tre. That compression will transfer to the neighbouring part of the rock, and so on through to the far side of the rock, from where it will bounce back to the top — all in a fraction of a second. This is known as a compression wave, and it can be illustrated by holding a loose spring (like a Slinky) that is attached to something (or someone) at the other end. If you give it a sharp push so the coils are compressed, the compression propagates (travels) along the length of the spring and back. You can think of a compression wave as a "push" wave — it's called a P-wave (although the "P" stands for "primary" because P-waves arrive first at seismic stations).

Hitting a large block of rock with a heavy hammer will create seismic waves within the rock.

When we hit a rock with a hammer, we also create a different type of body wave, one that is charac-terized by back-and-forth vibrations (as opposed to compressions). This is known as a shear wave

(S-wave, where the "S" stands for "secondary"), and an analogy would be what happens when you flick a length of rope with an up-and-down motion. A wave will form in the rope, which will travel to the end of the rope and back.

Compression waves and shear waves travel very quickly through geological materials. Typical P-wave velocities are between 0.5 km/s and 2.5 km/s in unconsolidated sediments, and between 3.0 km/s and 6.5 km/s in solid crustal rocks. Of the common rocks of the crust, velocities are greatest in basalt and granite. S-waves are slower than P-waves, with velocities between 0.1 km/s and 0.8 km/s in soft sediments, and between 1.5 km/s and 3.8 km/s in solid rocks.

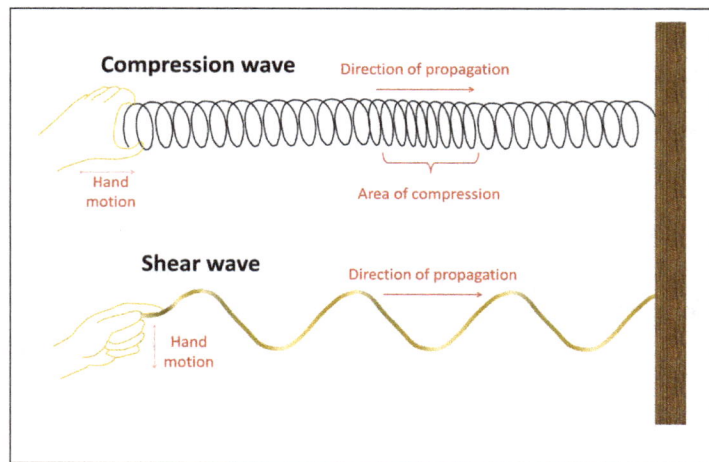

A compression wave can be illustrated by a spring (like a Slinky) that is given a sharp push at one end. A shear wave can be illustrated by a rope that is given a quick flick.

Mantle rock is generally denser and stronger than crustal rock and both P- and S-waves travel faster through the mantle than they do through the crust. Moreover, seismic-wave velocities are related to how tightly compressed a rock is, and the level of compression increases dramatically with depth. Finally, seismic waves are affected by the phase state of rock. They are slowed if there is any degree of melting in the rock. If the material is completely liquid, P-waves are slowed dramatically and S-waves are stopped altogether.

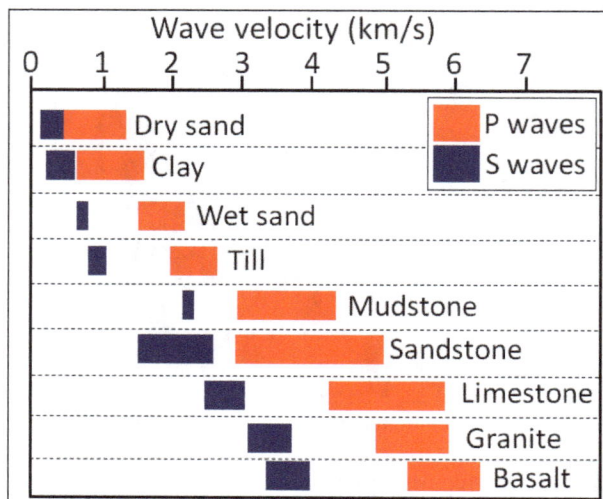

Typical velocities of P-waves (red) and S-waves (blue) in sediments and in solid crustal rocks.

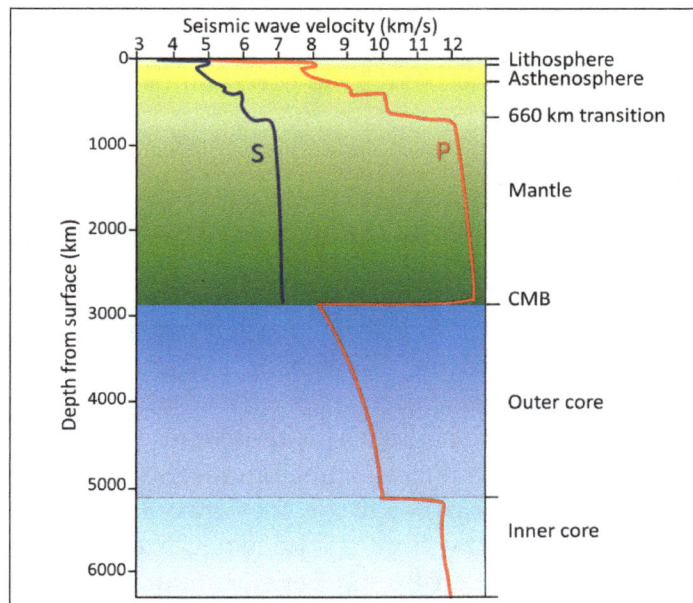

P-wave and S-wave velocity variations with depth in Earth.

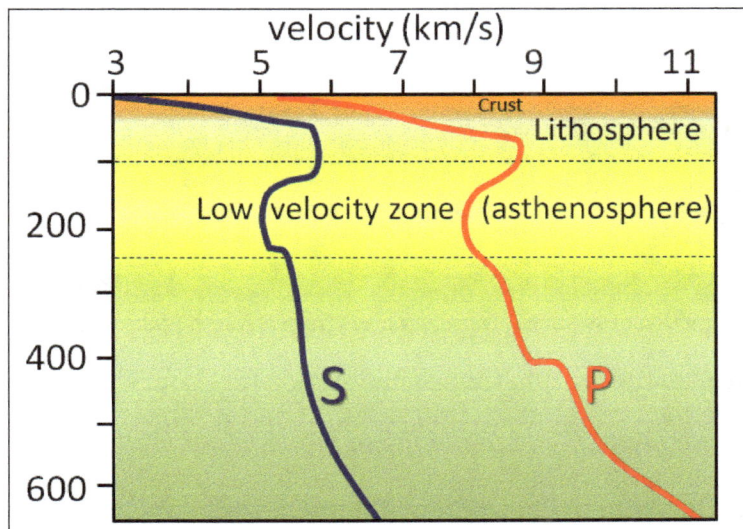

P-wave and S-wave velocity variations in the upper mantle and crust.

Accurate seismometers have been used for earthquake studies since the late 1800s, and systematic use of seismic data to understand Earth's interior started in the early 1900s. The rate of change of seismic waves with depth in Earth has been determined over the past several decades by analyzing seismic signals from large earthquakes at seismic stations around the world. Small differences in arrival time of signals at different locations have been interpreted to show that:

- Velocities are greater in mantle rock than in the crust.

- Velocities generally increase with pressure, and therefore with depth.

- Velocities slow in the area between 100 km and 250 km depth (called the "low-velocity zone"; equivalent to the asthenosphere).

- Velocities increase dramatically at 660 km depth (because of a mineralogical transition).

- Velocities slow in the region just above the core-mantle boundary (the D" layer or "ultra-low-velocity zone").

- S-waves do not pass through the outer part of the core.

- P-wave velocities increase dramatically at the boundary between the liquid outer core and the solid inner core.

One of the first discoveries about Earth's interior made through seismology was in the early 1900s when Croatian seismologist Andrija Mohorovičić realized that at certain distances from an earthquake, two separate sets of seismic waves arrived at a seismic station within a few seconds of each other. He reasoned that the waves that went down into the mantle, travelled through the mantle, and then were bent upward back into the crust, reached the seismic station first because although they had farther to go, they travelled faster through mantle rock. The boundary between the crust and the mantle is known as the Mohorovičić discontinuity (or Moho). Its depth is between 60 km and 80 km beneath major mountain ranges, around 30 km to 50 km beneath most of the continental crust, and between 5 km and 10 km beneath the oceanic crust.

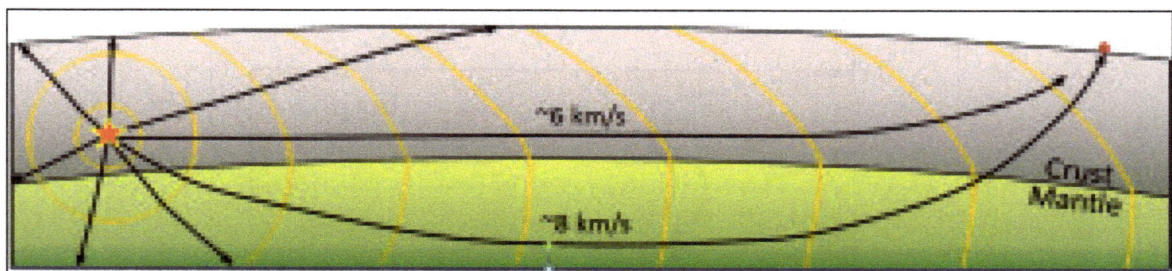

Depiction of seismic waves emanating from an earthquake (red star).

Some waves travel through the crust to the seismic station (at about 6 km/s), while others go down into the mantle (where they travel at around 8 km/s) and are bent upward toward the surface, reaching the station before the ones that travelled only through the crust.

Because of the gradual increase in density (and therefore rock strength) with depth, all waves are refracted (toward the lower density material) as they travel through homogenous parts of Earth and thus tend to curve outward toward the surface. Waves are also refracted at boundaries within Earth, such as at the Moho, at the core-mantle boundary (CMB), and at the outer-core/inner-core boundary.

S-waves do not travel through liquids — they are stopped at the CMB — and there is an S-wave shadow on the side of Earth opposite a seismic source. The angular distance from the seismic source to the shadow zone is 103° on either side, so the total angular distance of the shadow zone is 154°. We can use this information to infer the depth to the CMB.

P-waves do travel through liquids, so they can make it through the liquid part of the core. Because of the refraction that takes place at the CMB, waves that travel through the core are bent away from the surface, and this creates a P-wave shadow zone on either side, from 103° to 150°. This information can be used to discover the differences between the inner and outer parts of the core.

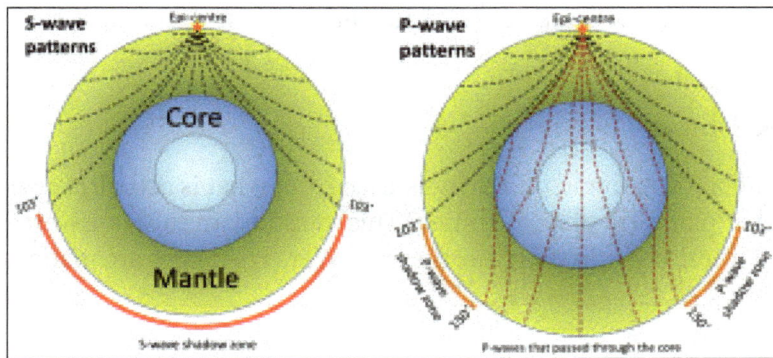

Patterns of seismic wave propagation through Earth's mantle and core. S-waves do not travel through the liquid outer core, so they leave a shadow on Earth's far side. P-waves do travel through the core, but because the waves that enter the core are refracted, there are also P-wave shadow zones.

Using data from many seismometers and hundreds of earthquakes, it is possible to create a two- or three-dimensional image of the seismic properties of part of the mantle. This technique is known as seismic tomography.

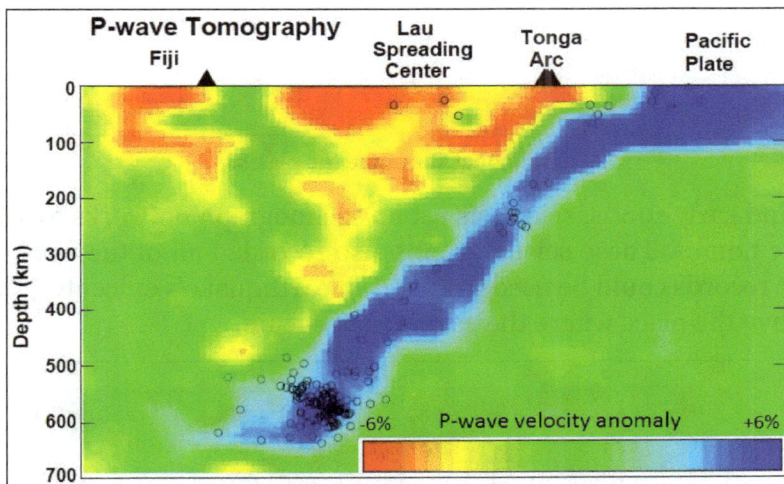

P-wave tomographic profile of area in the southern Pacific Ocean from southeast of Tonga to Fiji. Blue represents rock that has relatively high seismic velocities, while yellow and red represent rock with low velocities. Open circles are earthquakes used in the study.

The Pacific Plate subducts beneath Tonga and appears as a 100 km thick slab of cold (blue-coloured) oceanic crust that has pushed down into the surrounding hot mantle. The cold rock is more rigid than the surrounding hot mantle rock, so it is characterized by slightly faster seismic velocities. There is volcanism in the Lau spreading centre and also in the Fiji area, and the warm rock in these areas has slower seismic velocities (yellow and red colours).

THEORY OF PLATE TECTONICS

When the concept of seafloor spreading came along, scientists recognized that it was the mechanism to explain how continents could move around Earth's surface. Like the scientists before us, we will now merge the ideas of continental drift and seafloor spreading into the theory of plate tectonics.

Earth's Tectonic Plates

Seafloor and continents move around on Earth's surface, but what is actually moving? What portion of the Earth makes up the "plates" in plate tectonics? This question was also answered because of technology developed during war times – in this case, the Cold War. The plates are made up of the lithosphere.

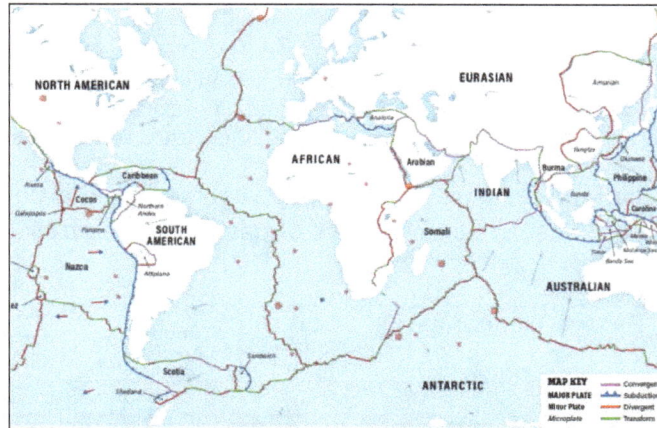

Earthquakes outline the plates.

During the 1950s and early 1960s, scientists set up seismograph networks to see if enemy nations were testing atomic bombs. These seismographs also recorded all of the earthquakes around the planet. The seismic records could be used to locate an earthquake's epicenter, the point on Earth's surface directly above the place where the earthquake occurs.

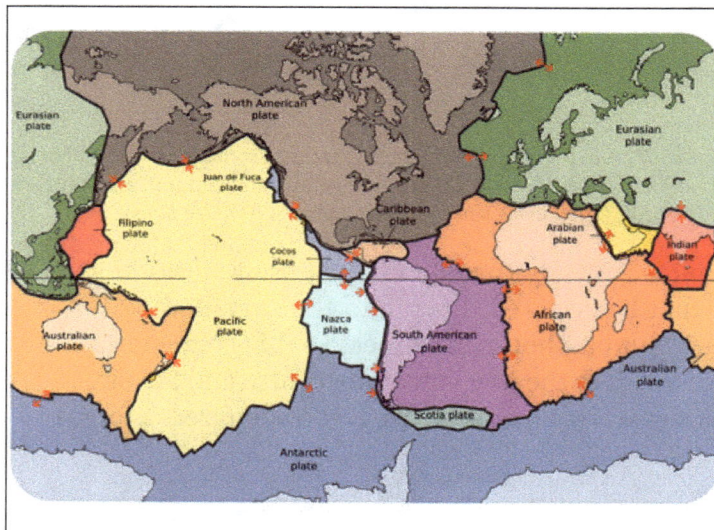

The lithospheric plates and their names. The arrows show whether the plates
are moving apart, moving together, or sliding past each other.

Earthquake epicenters outline the plates. Mid-ocean ridges, trenches, and large faults mark the edges of the plates, and this is where earthquakes occur.

The lithosphere is divided into a dozen major and several minor plates. The plates' edges can be drawn by connecting the dots that mark earthquakes' epicenters. A single plate can be made of all oceanic lithosphere or all continental lithosphere, but nearly all plates are made of a combination of both.

Movement of the plates over Earth's surface is termed plate tectonics. Plates move at a rate of a few centimeters a year, about the same rate fingernails grow.

How Plates Move

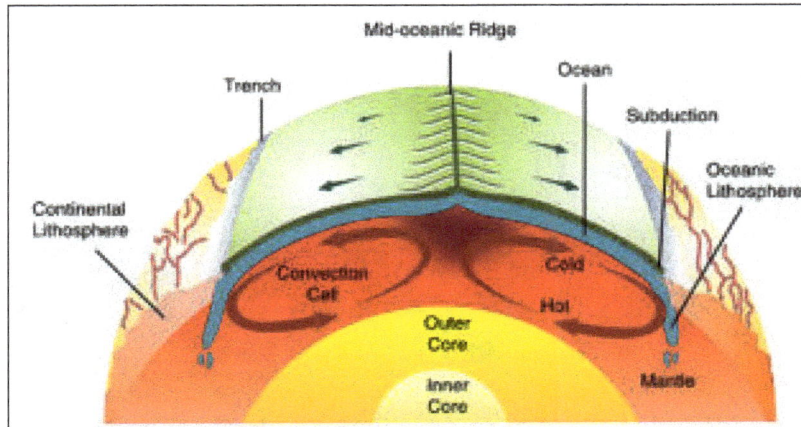

Mantle convection drives plate tectonics. Hot material rises at mid-ocean ridges and sinks at deep sea trenches, which keeps the plates moving along the Earth's surface.

If seafloor spreading drives the plates, what drives seafloor spreading? Picture two convection cells side-by-side in the mantle.

- Hot mantle from the two adjacent cells rises at the ridge axis, creating new ocean crust.

- The top limb of the convection cell moves horizontally away from the ridge crest, as does the new seafloor.

- The outer limbs of the convection cells plunge down into the deeper mantle, dragging oceanic crust as well. This takes place at the deep sea trenches.

- The material sinks to the core and moves horizontally.

- The material heats up and reaches the zone where it rises again.

Plate Boundaries

Plate boundaries are the edges where two plates meet. Most geologic activities, including volcanoes, earthquakes, and mountain building, take place at plate boundaries. How can two plates move relative to each other?

- Divergent plate boundaries: The two plates move away from each other.

- Convergent plate boundaries: The two plates move towards each other.

- Transform plate boundaries: The two plates slip past each other.

The type of plate boundary and the type of crust found on each side of the boundary determines what sort of geologic activity will be found there.

Divergent Plate Boundaries

Plates move apart at mid-ocean ridges where new seafloor forms. Between the two plates is a rift valley. Lava flows at the surface cool rapidly to become basalt, but deeper in the crust, magma cools more slowly to form gabbro. So the entire ridge system is made up of igneous rock that is either extrusive or intrusive. Earthquakes are common at mid-ocean ridges since the movement of magma and oceanic crust results in crustal shaking. The vast majority of mid-ocean ridges are located deep below the sea.

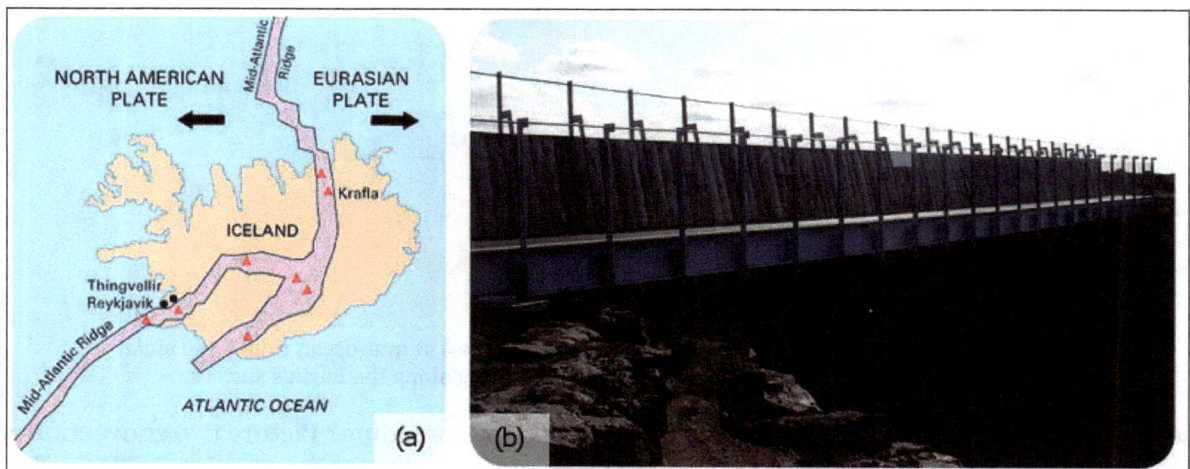

(a) Iceland is the one location where the ridge is located on land: the Mid-Atlantic Ridge separates the North American and Eurasian plates; (b) The rift valley in the Mid-Atlantic Ridge on Iceland.

The Arabian, Indian, and African plates are rifting apart, forming the Great Rift Valley in Africa. The Dead Sea fills the rift with seawater.

Can divergent plate boundaries occur within a continent? What is the result? Incontinental rifting, magma rises beneath the continent, causing it to become thinner, break, and ultimately split apart. New ocean crust erupts in the void, creating an ocean between continents.

Convergent Plate Boundaries

When two plates converge, the result depends on the type of lithosphere the plates are made of. No matter what, smashing two enormous slabs of lithosphere together results in magma generation and earthquakes.

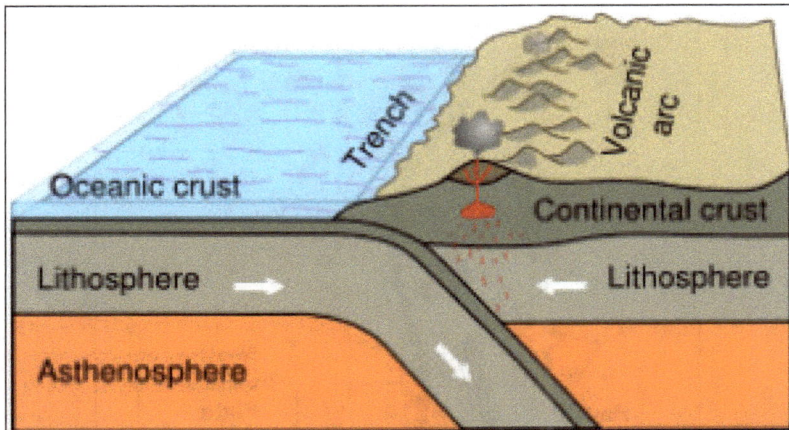

Subduction of an oceanic plate beneath a continental plate causes earthquakes and forms a line of volcanoes known as a continental arc.

Ocean-continent

When oceanic crust converges with continental crust, the denser oceanic plate plunges beneath the continental plate. This process, called subduction, occurs at the oceanic trenches. The entire region is known as a subduction zone. Subduction zones have a lot of intense earthquakes and volcanic eruptions. The subducting plate causes melting in the mantle. The magma rises and erupts, creating volcanoes. These coastal volcanic mountains are found in a line above the subducting plate. The volcanoes are known as a continental arc.

(a) At the trench lining the western margin of South America, the Nazca plate is subducting beneath the South American plate, resulting in the Andes Mountains (brown and red uplands); (b) Convergence has pushed up limestone in the Andes Mountains where volcanoes are common.

The movement of crust and magma causes earthquakes. The volcanoes of northeastern California — Lassen Peak, Mount Shasta, and Medicine Lake volcano — along with the rest of the Cascade Mountains of the Pacific Northwest are the result of subduction of the Juan de Fuca plate beneath the North American plate . The Juan de Fuca plate is created by seafloor spreading just offshore at the Juan de Fuca ridge.

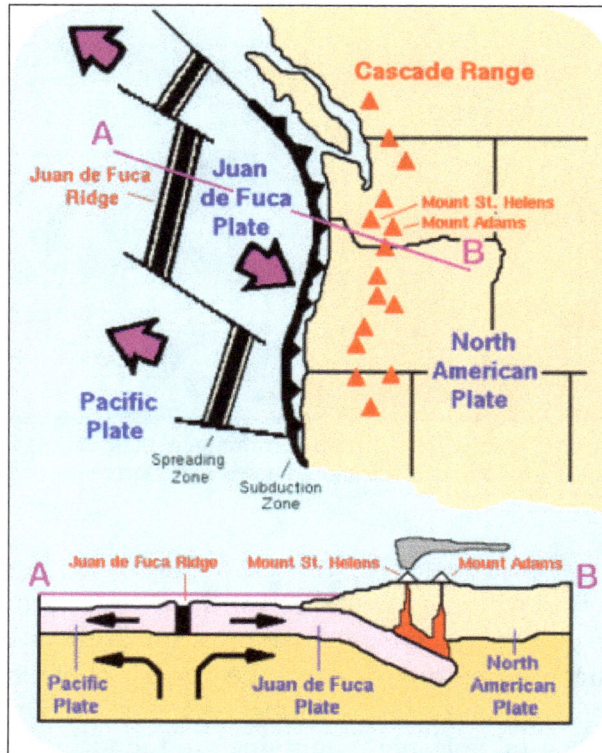

The Cascade Mountains of the Pacific Northwest are a continental arc.

If the magma at a continental arc is felsic, it may be too viscous (thick) to rise through the crust. The magma will cool slowly to form granite or granodiorite. These large bodies of intrusive igneous rocks are called batholiths, which may someday be uplifted to form a mountain range.

The Sierra Nevada batholith cooled beneath a volcanic arc roughly 200 million years ago. The rock is well exposed here at Mount Whitney. Similar batholiths are likely forming beneath the Andes and Cascades today.

Ocean-ocean

When two oceanic plates converge, the older, denser plate will subduct into the mantle. An ocean trench marks the location where the plate is pushed down into the mantle. The line of volcanoes that grows on the upper oceanic plate is an island arc.

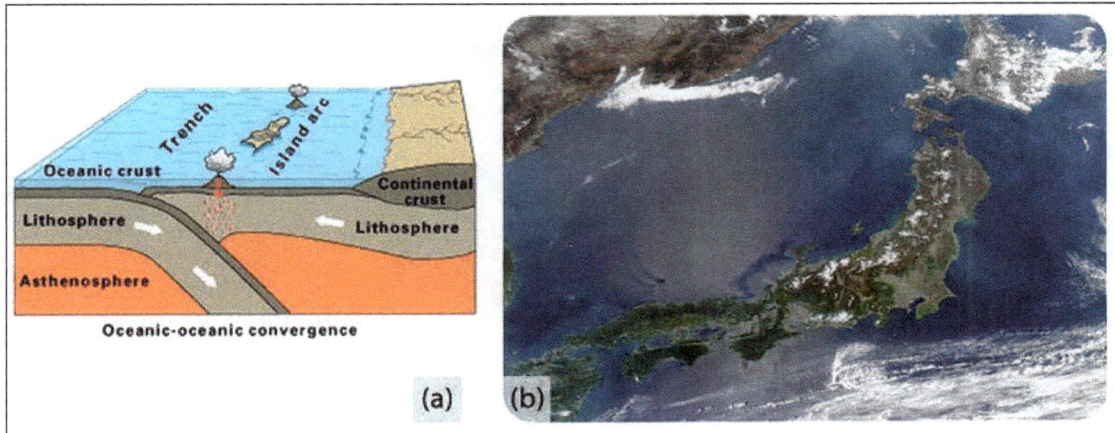

(a) Subduction of an ocean plate beneath an ocean plate results in a volcanic island arc, an ocean trench and many earthquakes. (b) Japan is an arc-shaped island arc composed of volcanoes off the Asian mainland, as seen in this satellite image.

Continent-continent

Continental plates are too buoyant to subduct. What happens to continental material when it collides? Since it has nowhere to go but up, this creates some of the world's largest mountains ranges. Magma cannot penetrate this thick crust so there are no volcanoes, although the magma stays in the crust. Metamorphic rocks are common because of the stress the continental crust experiences. With enormous slabs of crust smashing together, continent-continent collisions bring on numerous and large earthquakes.

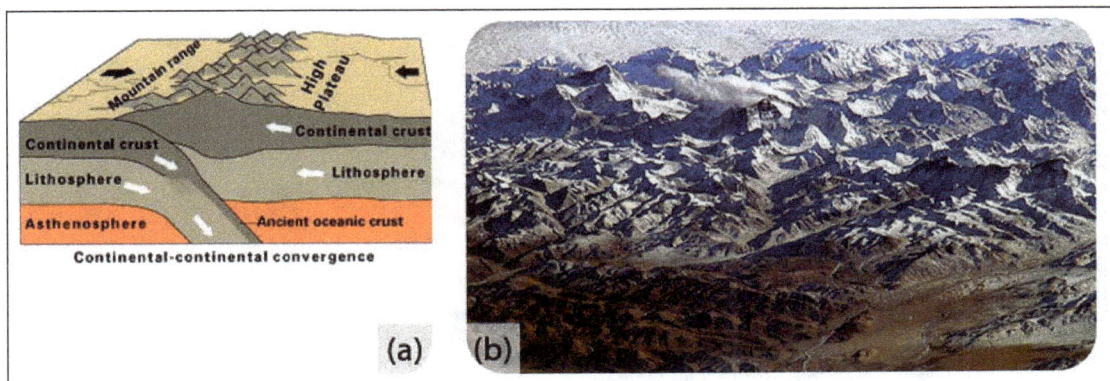

(a) In continent-continent convergence, the plates push upward to create a high mountain range. (b) The world's highest mountains, the Himalayas, are the result of the collision of the Indian Plate with the Eurasian Plate, seen in this photo from the International Space Station.

The Appalachian Mountains are the remnants of a large mountain range that was created when North America rammed into Eurasia about 250 million years ago.

Transform Plate Boundaries

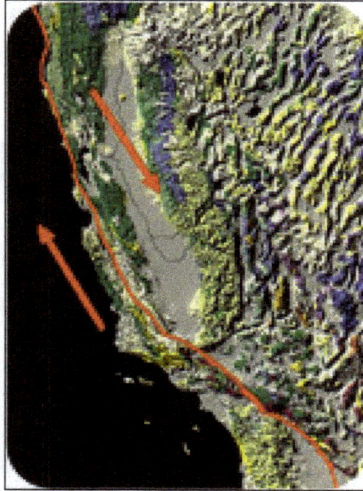

At the San Andreas Fault in California, the Pacific Plate is sliding northwest relative to the North American plate, which is moving southeast. At the northern end of the picture, the transform boundary turns into a subduction zone.

Transform plate boundaries are seen as transform faults, where two plates move past each other in opposite directions. Transform faults on continents bring massive earthquakes.

California is very geologically active. What are the three major plate boundaries in or near California?

- A transform plate boundary between the Pacific and North American plates creates the San Andreas Fault, the world's most notorious transform fault.

- Just offshore, a divergent plate boundary, Juan de Fuca ridge, creates the Juan de Fuca plate.

- A convergent plate boundary between the Juan de Fuca oceanic plate and the North American continental plate creates the Cascades volcanoes.

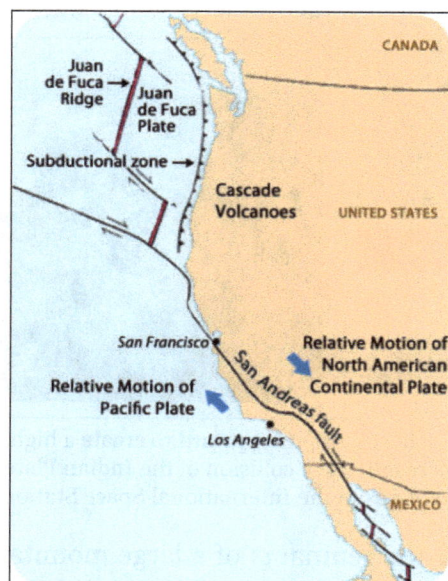

This map shows the three major plate boundaries in or near California.

Earth's Changing Surface

Geologists know that Wegener was right because the movements of continents explain so much about the geology we see. Most of the geologic activity that we see on the planet today is because of the interactions of the moving plates.

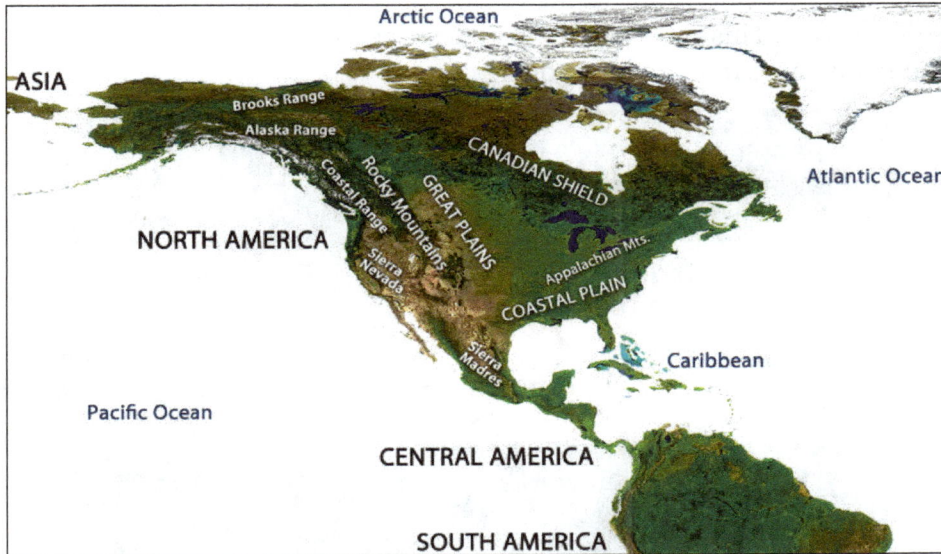

Mountain ranges of North America.

In the map of North America, where are the mountain ranges located? Using what you have learned about plate tectonics, try to answer the following questions:

- What is the geologic origin of the Cascades Range? The Cascades are a chain of volcanoes in the Pacific Northwest. They are not labelled on the diagram but they lie between the Sierra Nevada and the Coastal Range.

- What is the geologic origin of the Sierra Nevada? (These mountains are made of granitic intrusions.)

- What is the geologic origin of the Appalachian Mountains along the Eastern US?

About 200 million years ago, the Appalachian Mountains of eastern North America were probably once as high as the Himalaya, but they have been weathered and eroded significantly since the breakup of Pangaea.

Remember that Wegener used the similarity of the mountains on the west and east sides of the Atlantic as evidence for his continental drift hypothesis. The Appalachian mountains formed at a convergent plate boundary as Pangaea came togethe.

Before Pangaea came together, the continents were separated by an ocean where the Atlantic is now. The proto-Atlantic ocean shrank as the Pacific Ocean grew. Currently, the Pacific is shrinking as the Atlantic is growing. This supercontinent cycle is responsible for most of the geologic features that we see and many more that are long gone.

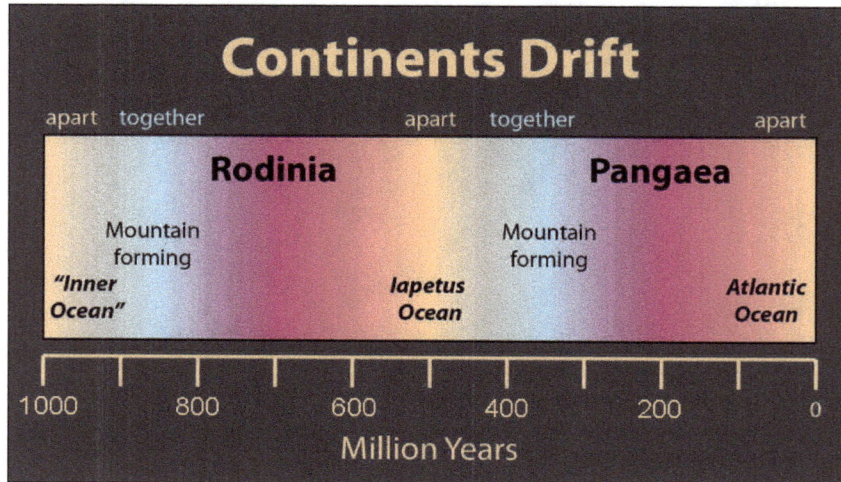

Scientists think that the creation and breakup of a supercontinent takes place about every 500 million years. The supercontinent before Pangaea was Rodinia. A new continent will form as the Pacific ocean disappears.

Developing the Theory

In line with other previous and contemporaneous proposals, in 1912 the meteorologist Alfred Wegener amply described what he called continental drift, and the scientific debate started that would end up fifty years later in the theory of plate tectonics. Starting from the idea (also expressed by his forerunners) that the present continents once formed a single land mass (which was called Pangea later on) that drifted apart, thus releasing the continents from the Earth's mantle and likening them to "icebergs" of low density granite floating on a sea of denser basalt.

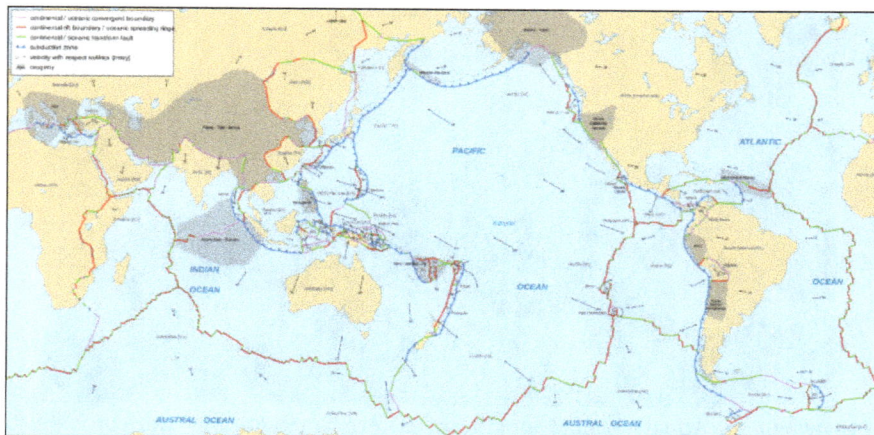

Detailed map showing the tectonic plates with their movement vectors.

Supporting evidence for the idea came from the dove-tailing outlines of South America's east coast and Africa's west coast, and from the matching of the rock formations along these edges. Confirmation of their previous contiguous nature also came from the fossil plants Glossopteris and Gangamopteris, and the therapsid or mammal-like reptile Lystrosaurus, all widely distributed over South America, Africa, Antarctica, India and Australia. The evidence for such an erstwhile joining of these continents was patent to field geologists working in the southern hemisphere. The South African Alex du Toit put together a mass of such information, and went further than Wegener in recognising the strong links between the Gondwana fragments.

But without detailed evidence and a force sufficient to drive the movement, the theory was not generally accepted: The Earth might have a solid crust and mantle and a liquid core, but there seemed to be no way that portions of the crust could move around. Distinguished scientists, such as Harold Jeffreys and Charles Schuchert, were outspoken critics of continental drift.

Despite much opposition, the view of continental drift gained support and a lively debate started between "drifters" or "mobilists" (proponents of the theory) and "fixists" (opponents). During the 1920s, 1930s and 1940s, the former reached important milestones proposing that convection currents might have driven the plate movements, and that spreading may have occurred below the sea within the oceanic crust. Concepts close to the elements now incorporated in plate tectonics were proposed by geophysicists and geologists (both fixists and mobilists) like Vening-Meinesz, Holmes, and Umbgrove.

One of the first pieces of geophysical evidence that was used to support the movement of lithospheric plates came from paleomagnetism. This is based on the fact that rocks of different ages show a variable magnetic field direction, evidenced by studies since the mid-nineteenth century. The magnetic north and south poles reverse through time, and, especially important in paleotectonic studies, the relative position of the magnetic north pole varies through time. Initially, during the first half of the twentieth century, the latter phenomenon was explained by introducing what was called "polar wander", i.e. it was assumed that the North Pole location had been shifting through time.

An alternative explanation, though, was that the continents had moved (shifted and rotated) relative to the North Pole, and each continent, in fact, shows its own "polar wander path". During the late 1950s, it was successfully shown on two occasions that these data could show the validity of continental drift: By Keith Runcorn in a paper in 1956, and by Warren Carey in a symposium held in March 1956.

The second piece of evidence in support of continental drift came during the late 1950s and early 60s from data on the bathymetry of the deep ocean floors and the nature of the oceanic crust such as magnetic properties and, more generally, with the development of marine geology which gave evidence for the association of seafloor spreading along the mid-oceanic ridges and magnetic field reversals.

Simultaneous advances in early seismic imaging techniques in and around Wadati-Benioff zones along the trenches bounding many continental margins, together with many other geophysical (e.g. gravimetric) and geological observations, showed how the oceanic crust could disappear into the mantle, providing the mechanism to balance the extension of the ocean basins with shortening along its margins.

All this evidence, both from the ocean floor and from the continental margins, made it clear around 1965 that continental drift was feasible and the theory of plate tectonics, which was defined in a series of papers between 1965 and 1967, was born, with all its extraordinary explanatory and predictive power. The theory revolutionized the Earth sciences, explaining a diverse range of geological phenomena and their implications in other studies such as paleogeography and paleobiology.

Continental Drift

Alfred Wegener.

In the late nineteenth and early twentieth centuries, geologists assumed that the Earth's major features were fixed, and that most geologic features such as basin development and mountain ranges could be explained by vertical crustal movement, described in what is called the geosynclinal theory. Generally, this was placed in the context of a contracting planet Earth due to heat loss in the course of a relatively short geological time.

It was observed as early as 1596 that the opposite coasts of the Atlantic Ocean — or, more precisely, the edges of the continental shelves — have similar shapes and seem to have once fitted together.

Since that time many theories were proposed to explain this apparent complementarity, but the assumption of a solid Earth made these various proposals difficult to accept.

The discovery of radioactivity and its associated heating properties in 1895 prompted a re-examination of the apparent age of the Earth. This had previously been estimated by its cooling rate and assumption the Earth's surface radiated like a black body. Those calculations had implied that, even if it started at red heat, the Earth would have dropped to its present temperature in a few tens of millions of years. Armed with the knowledge of a new heat source, scientists realized that the Earth would be much older, and that its core was still sufficiently hot to be liquid.

Floating Continents, Paleomagnetism and Seismicity Zones

As it was observed early that although granite existed on continents, seafloor seemed to be composed of denser basalt, the prevailing concept during the first half of the twentieth century was that there were two types of crust, named "sial" (continental type crust) and "sima" (oceanic type crust). Furthermore, it was supposed that a static shell of strata was present under the continents. It therefore looked apparent that a layer of basalt (sial) underlies the continental rocks.

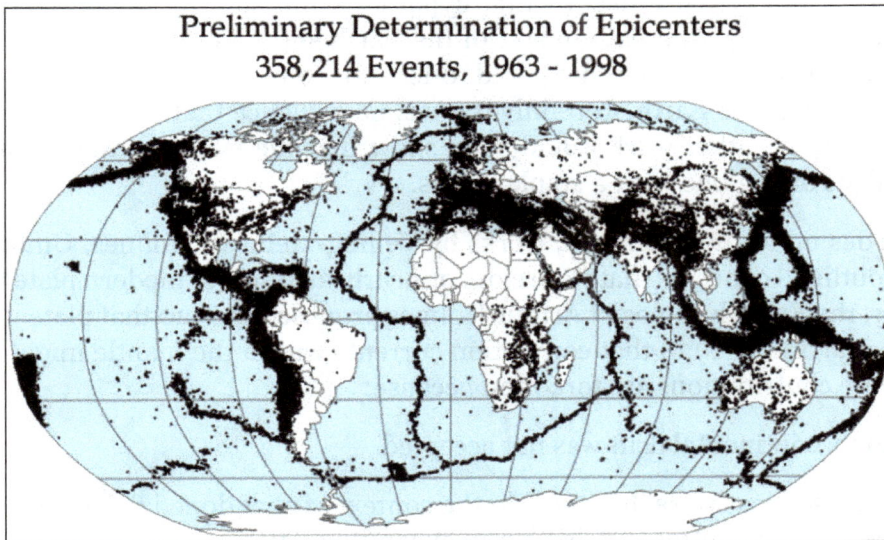

Global earthquake epicenters.

However, based on abnormalities in plumb line deflection by the Andes in Peru, Pierre Bouguer had deduced that less-dense mountains must have a downward projection into the denser layer underneath. The concept that mountains had "roots" was confirmed by George B. Airy a hundred years later, during study of Himalayan gravitation, and seismic studies detected corresponding density variations. Therefore, by the mid-1950s, the question remained unresolved as to whether mountain roots were clenched in surrounding basalt or were floating on it like an iceberg.

During the twentieth century, improvements in and greater use of seismic instruments such as seismographs enabled scientists to learn that earthquakes tend to be concentrated in specific areas, most notably along the oceanic trenches and spreading ridges. By the late 1920s, seismologists were beginning to identify several prominent earthquake zones parallel to the trenches that typically were inclined 40–60° from the horizontal and extended several hundred kilometers into the Earth. These zones later became known as Wadati-Benioff zones, or simply Benioff zones, in honor of the seismologists who first recognized them, Kiyoo Wadati of Japan and Hugo Benioff of the United States. The study of global seismicity greatly advanced in the 1960s with the establishment of the Worldwide Standardized Seismograph Network (WWSSN) to monitor the compliance of the 1963 treaty banning above-ground testing of nuclear weapons. The much improved data from the WWSSN instruments allowed seismologists to map precisely the zones of earthquake concentration worldwide.

Meanwhile, debates developed around the phenomena of polar wander. Since the early debates of continental drift, scientists had discussed and used evidence that polar drift had occurred because continents seemed to have moved through different climatic zones during the past. Furthermore, paleomagnetic data had shown that the magnetic pole had also shifted during time. Reasoning in an opposite way, the continents might have shifted and rotated, while the pole remained relatively fixed. The first time the evidence of magnetic polar wander was used to support the movements of continents by Keith Runcorn in 1956. This was immediately followed by a symposium in Tasmania in March 1956. In this symposium, the evidence was used in the theory of an expansion of the global crust. In this hypothesis the shifting of the continents can be simply explained by a large increase in size of the Earth since its formation. However, this was unsatisfactory because its supporters could offer no convincing mechanism to produce a significant expansion of the Earth. Certainly there is no evidence that the moon has expanded in the past 3 billion years; other work would soon show that the evidence was equally in support of continental drift on a globe with a stable radius.

During the thirties up to the late fifties, works by Vening-Meinesz, Holmes, Umbgrove, and numerous others outlined concepts that were close or nearly identical to modern plate tectonics theory. In particular, the English geologist Arthur Holmes proposed in 1920 that plate junctions might lie beneath the sea, and in 1928 that convection currents within the mantle might be the driving force. Often, these contributions are forgotten because:

- At the time, continental drift was not accepted.

- Some of these ideas were discussed in the context of abandoned fixistic ideas of a deforming globe without continental drift or an expanding Earth.

- They were published during an episode of extreme political and economic instability that hampered scientific communication.

- Many were published by European scientists and at first not mentioned or given little credit in the papers on sea floor spreading published by the American researchers in the 1960s.

Mid-oceanic Ridge Spreading and Convection

In 1947, a team of scientists led by Maurice Ewing utilizing the Woods Hole Oceanographic Institution's research vessel Atlantis and an array of instruments, confirmed the existence of a rise in the central Atlantic Ocean, and found that the floor of the seabed beneath the layer of sediments consisted of basalt, not the granite which is the main constituent of continents. They also found that the oceanic crust was much thinner than continental crust. All these new findings raised important and intriguing questions.

The new data that had been collected on the ocean basins also showed particular characteristics regarding the bathymetry. One of the major outcomes of these datasets was that all along the globe, a system of mid-oceanic ridges was detected. An important conclusion was that along this system, new ocean floor was being created, which led to the concept of the "Great Global Rift". This was described in the crucial paper of Bruce Heezen, which would trigger a real revolution in thinking. A profound consequence of seafloor spreading is that new crust was, and still is, being continually created along the oceanic ridges. Therefore, Heezen advocated the so-called "expanding Earth" hypothesis of S. Warren Carey. So, still the question remained: How can new crust be continuously added along the oceanic ridges without increasing the size of the Earth? In reality, this question had been solved already by numerous scientists during the forties and the fifties, like Arthur Holmes, Vening-Meinesz, Coates and many others: The crust in excess disappeared along what were called the oceanic trenches, where so-called "subduction" occurred. Therefore, when various scientists during the early sixties started to reason on the data at their disposal regarding the ocean floor, the pieces of the theory quickly fell into place.

The question particularly intrigued Harry Hammond Hess, a Princeton University geologist and a Naval Reserve Rear Admiral, and Robert S. Dietz, a scientist with the U.S. Coast and Geodetic Survey who first coined the term seafloor spreading. Dietz and Hess were among the small handful who really understood the broad implications of sea floor spreading and how it would eventually agree with the, at that time, unconventional and unaccepted ideas of continental drift and the elegant and mobilistic models proposed by previous workers like Holmes.

In the same year, Robert R. Coats of the U.S. Geological Survey described the main features of island arc subduction in the Aleutian Islands. His paper, though little noted (and even ridiculed) at the time, has since been called "seminal" and "prescient." In reality, it actually shows that the work by the European scientists on island arcs and mountain belts performed and published during the 1930s up until the 1950s was applied and appreciated also in the United States.

If the Earth's crust was expanding along the oceanic ridges, Hess and Dietz reasoned like Holmes and others before them, it must be shrinking elsewhere. Hess followed Heezen, suggesting that new oceanic crust continuously spreads away from the ridges in a conveyor belt – like motion. And, using the mobilistic concepts developed before, he correctly concluded that many millions of years later, the oceanic crust eventually descends along the continental margins where oceanic trenches — very deep, narrow canyons — are formed, e.g. along the rim of the Pacific Ocean basin. The important step Hess made was that convection currents would be the driving force in this process, arriving at the same conclusions as Holmes had decades before with the only difference that the thinning of the ocean crust was performed using Heezen's mechanism of spreading along the ridges. Hess therefore concluded that the Atlantic Ocean was expanding while the Pacific Ocean was shrinking. As old oceanic crust is "consumed" in the trenches (like Holmes and others, he thought this was done by thickening of the continental lithosphere, not, as now understood, by underthrusting at a larger scale of the oceanic crust itself into the mantle), new magma rises and erupts along the spreading ridges to form new crust. In effect, the ocean basins are perpetually being "recycled," with the creation of new crust and the destruction of old oceanic lithosphere occurring simultaneously. Thus, the new mobilistic concepts neatly explained why the Earth does not get bigger with sea floor spreading, why there is so little sediment accumulation on the ocean floor, and why oceanic rocks are much younger than continental rocks.

Magnetic Striping

Beginning in the 1950s, scientists like Victor Vacquier, using magnetic instruments (magnetometers) adapted from airborne devices developed during World War II to detect submarines, began recognizing odd magnetic variations across the ocean floor. This finding, though unexpected, was not entirely surprising because it was known that basalt — the iron-rich, volcanic rock making up the ocean floor — contains a strongly magnetic mineral (magnetite) and can locally distort compass readings. This distortion was recognized by Icelandic mariners as early as the late eighteenth century. More important, because the presence of magnetite gives the basalt measurable magnetic properties, these newly discovered magnetic variations provided another means to study the deep ocean floor. When newly formed rock cools, such magnetic materials recorded the Earth's magnetic field at the time.

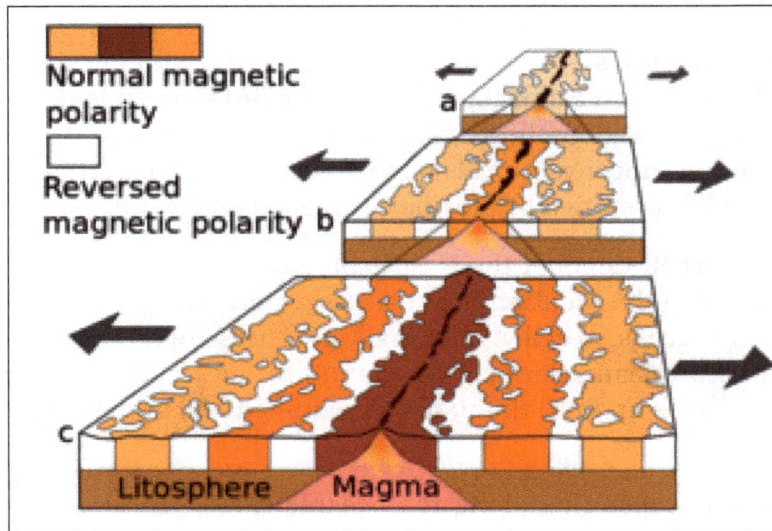

Seafloor magnetic striping.

As more and more of the seafloor was mapped during the 1950s, the magnetic variations turned out not to be random or isolated occurrences, but instead revealed recognizable patterns. When these magnetic patterns were mapped over a wide region, the ocean floor showed a zebra-like pattern: One stripe with normal polarity and the adjoining stripe with reversed polarity. The overall pattern, defined by these alternating bands of normally and reversely polarized rock, became known as magnetic striping, and was published by Ron G. Mason and co-workers in 1961, who did not find, though, an explanation for these data in terms of sea floor spreading, like Vine, Matthews and Morley a few years later.

A demonstration of magnetic striping. The darker the color is, the closer it is to normal polarity.

The discovery of magnetic striping called for an explanation. In the early 1960s scientists such as Heezen, Hess and Dietz had begun to theorise that mid-ocean ridges mark structurally weak zones where the ocean floor was being ripped in two lengthwise along the ridge crest.

New magma from deep within the Earth rises easily through these weak zones and eventually erupts along the crest of the ridges to create new oceanic crust. This process, at first denominated the "conveyer belt hypothesis" and later called seafloor spreading, operating over many millions of years continues to form new ocean floor all across the 50,000 km-long system of mid-ocean ridges.

Only four years after the maps with the "zebra pattern" of magnetic stripes were published, the link between sea floor spreading and these patterns was correctly placed, independently by Lawrence Morley, and by Fred Vine and Drummond Matthews, in 1963, now called the Vine-Matthews-Morley hypothesis. This hypothesis linked these patterns to geomagnetic reversals and was supported by several lines of evidence:

- The stripes are symmetrical around the crests of the mid-ocean ridges; at or near the crest of the ridge, the rocks are very young, and they become progressively older away from the ridge crest.

- The youngest rocks at the ridge crest always have present-day (normal) polarity.

- Stripes of rock parallel to the ridge crest alternate in magnetic polarity (normal-reversed-normal, etc.), suggesting that they were formed during different epochs documenting the (already known from independent studies) normal and reversal episodes of the Earth's magnetic field.

By explaining both the zebra-like magnetic striping and the construction of the mid-ocean ridge system, the seafloor spreading hypothesis (SFS) quickly gained converts and represented another major advance in the development of the plate-tectonics theory. Furthermore, the oceanic crust now came to be appreciated as a natural "tape recording" of the history of the geomagnetic field reversals (GMFR) of the Earth's magnetic field. Today, extensive studies are dedicated to the calibration of the normal-reversal patterns in the oceanic crust on one hand and known timescales derived from the dating of basalt layers in sedimentary sequences (magnetostratigraphy) on the other, to arrive at estimates of past spreading rates and plate reconstructions.

SEISMIC IMAGING AND INVERSION

The ability of seismic reflection technology to image subsurface targets is possible largely through the geometry of sources and receivers. A method similar to triangulation is used to place reflections in their correct locations with correct amplitudes, which can then be interpreted. The amplitudes are indicative of relative changes in impedance, and the seismic volume can be processed to yield impedances between the reflecting boundaries.

Stacking and Interval Velocities

The geometry of sources and receivers in a typical reflection seismic survey yields a number of seismic traces with common midpoints or central bins for stacking. These traces were recorded at different offset distances, and the travel times for seismic waves traveling to and from a given reflecting horizon varies with that distance. If the overburden through which the seismic waves pass is of constant velocity, then the time-variation with distance is a simple application of Pythagorean geometry, and the shape of the reflector on a seismic "gather" of traces is hyperbolic. As the overburden velocity structure becomes more complex, the shape is less perfectly hyperbolic, but most standard processing routines still assume a hyperbolic "moveout" of each reflector. An analysis is then made of selected seismic gathers to establish the ideal moveout required to "flatten" each reflection in the gather. This moveout is expressed in terms of a velocity and represents the seismic velocity that the entire overburden, down to the point of each particular reflection, would have to result in the idealized hyperbolic shape observed. This velocity analysis is usually conducted by examining the semblance (or some other measure of similarity) across all the traces, within a moving time window, and for all reasonable stacking velocities. The seismic processor then selects the best set of velocities to use at a variety of reflectors and constructs a velocity function of two-way travel time. These velocity functions are interpolated, both spatially and in two-way travel time, and all seismic gathers are then "corrected for normal moveout" using them. Each moveout-corrected gather is then summed or "stacked" after eliminating ("muting") those portions of the traces that have been highly distorted by the moveout process.

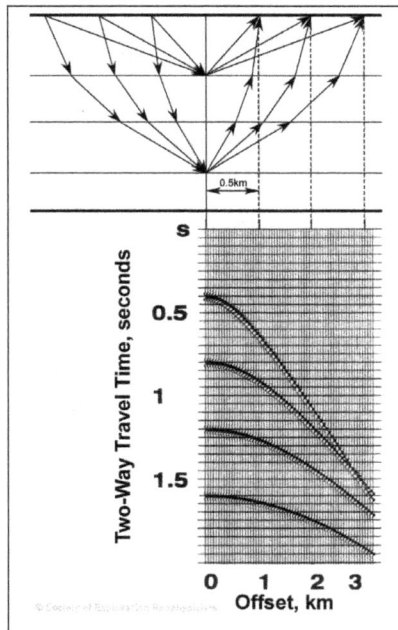

Ray diagram for normal moveout and a synthetic seismic gather. At the top of the figure is a schematic ray diagram, showing an earth model with four reflecting interfaces; rays are drawn from three source locations to three receiver locations, as they are reflected from two of the interfaces (the other source-receiver rays and reflections from other interfaces are not shown). The lower part of the figure shows the seismograms that would be recorded from this scenario, ignoring the direct wave in the upper layer, multiples, and noise. Notice that the distance used to label the seismic gathers is the total source receiver distance.

Velocity analysis on the right of a single common midpoint gather on the left. The gather is analyzed over narrow time windows for the values of semblance (or some other measure of similarity) according to a range of stacking velocities. The contours indicate the level of semblance, and the processing geophysicist selects the values deemed to be because of primary reflections and not events reflected multiple times. The direct waves (straight-line arrivals seen at the upper edge of the arrivals on the seismic gather) are not considered in the analysis.

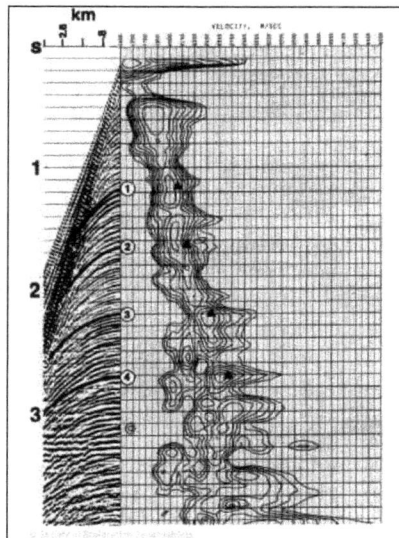

The final stacked traces exhibit a considerably better signal-to-noise ratio than the individual seismic traces recorded at zero-offset, but the improvement is better than just the square root of the number of traces that might be expected because of the systematic removal of coherent noise. Much of the noise present in individual seismic traces is not random but represents unwanted events, including surface waves or ground roll and multiply-reflected arrivals from shallow horizons; both of these can usually be identified in the velocity analysis and selected against. The stacking process then removes most of the unaligned energy associated with these types of coherent noise.

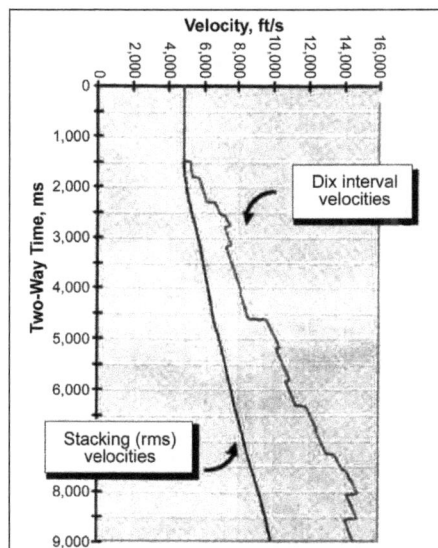

Interval velocities.

The velocities obtained in the analysis are not true seismic velocities — they are simply those velocities which provided the best stack of the data and may or may not truly reflect the actual root mean square (RMS) velocities that approximate the accumulated effect of the stack of layers above the reflector (the name RMS is derived from the arithmetic used to define this overall velocity). If we assume, however, that the stacking velocities do in fact provide a reasonable approximation to the aggregate effect of the layers overlying each reflector, the actual velocities of each layer can be obtained through a set of equations because of Dix. These "interval" or "Dix" velocities can sometimes be used to characterize the rocks in each layer and may be sufficiently precise to enable differentiation of gross rock types, although the errors associated with interval velocities can be fairly large.

Time and Depth Migration

Even after accounting for normal moveout and stacking the gathered traces to a common zero-offset equivalent set of traces, the locations of the reflected events are not usually correct because of lateral variations in velocity and dipping interfaces. Figure shows a simple 2D example of a dipping interface from which we observe a reflection. Each seismic trace is plotted directly beneath the respective midpoint or bin location used for stacking, but the reflection from any given interface may not have come from that location. The events have been shifted downdip to deeper locations, and the dip of the interface is less steep. To correct for this shift, the seismic processor "migrates" each sample to its appropriate position. In the simple case shown in the figure, we need only know the velocity of the one overlying layer, but in more realistic cases, the velocity function may be quite complex and is derived through a trial-and-error approach guided by statistical tests of lateral coherence, knowledge of expected geologic structure, and other constraints such as interval velocities and well log data. The problem can become quite difficult in complicated 3D data sets, and software has been developed to manage and visualize the velocity volume. The result of this model-driven 3D migration can be somewhat subjective, and, although it is possible to create structures where none really exist through this process, migration should be performed on all seismic data sets for appropriate imaging of structures. 3D migration can drastically improve the imaging of virtually any target by improving the accuracy of the spatial location of various features and by sharpening the image itself, allowing finer resolution than either migrated 2D data or unmigrated 3D data. The results can occasionally be quite dramatic for interpretation; for example, a locally high feature on an unmigrated data set may move to a significantly different map location after migration. In general, the more dramatic the structure, or the larger the velocity contrasts between layers, the more important 3D migration is for proper imaging.

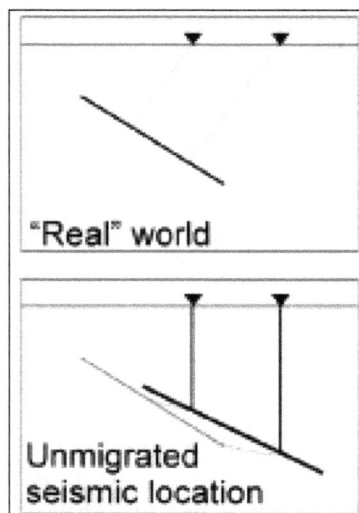

Migration of one dipping interface. The true Earth model (in two-way travel time) for a simple dipping interface is shown at the top, with normal-incidence (zero-offset) seismic rays drawn to two surface source-and-receiver locations. Because the seismograms are plotted directly beneath the surface locations, a seismic section will display the dipping interface at the incorrect location, as shown at the bottom. Notice that the seismic images the event downdip of its true location and with a less-steep dip. The processing step of migration attempts to correct for this, displacing the events back to their true locations.

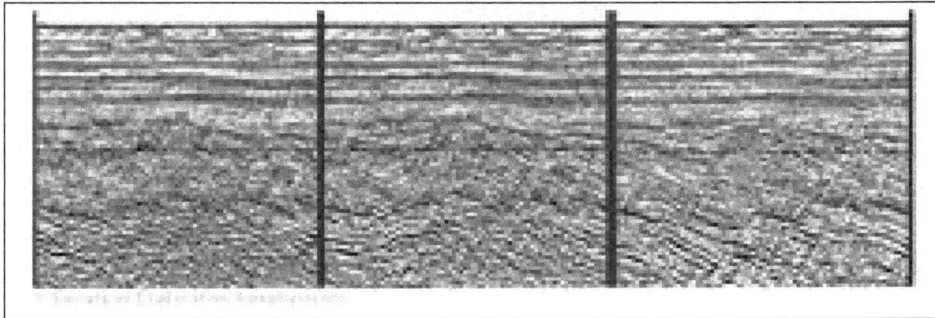

Example of improvement using 3D migration. These three panels show the same cross section of the Earth. The panel on the left was imaged as a 2D stack, extracted from a 3D data volume without migration. The panel in the center was imaged using 2D migration techniques. The panel on the right was imaged using 3D post-stack time migration.

The process of imaging through modeling the velocity structure is a form of inversion of seismic data, and the term inversion is often used to imply building a velocity model which is iteratively improved until it and the seismic data are optimally in agreement. Improvements in imaging are continually being made, and research in this area is one of the most fruitful in reservoir and exploration geophysics. The current methods of migration involve operating in two-way travel time, or in depth (using the model velocities to convert from travel time to depth), and either method can be performed prestack or post-stack. In addition, there have been a number of shortcuts developed over the years to provide reasonable results in a short time; all of the methods are quite computation-intensive, and the technology has benefited greatly from improved computing capacity. The finest results can usually be obtained from prestack depth migration, in which each sample of each trace, prior to gather, is migrated using the velocity function to a new location then stacked and compared with various tests for model improvement; the model is changed, and the process is repeated.

In areas where it is important to image beneath layers of high velocity contrasts, such as beneath salt bodies, prestack depth migration is required. The example shown in figure shows the possible improvements that can be obtained using prestack depth migration. The process required to create the final stack is as follows: a velocity model is first constructed through the water and sediment layers to the top of salt, and prestack depth migration is used to optimize that model. Then, the salt velocity (which is fairly constant and typically much higher than the surrounding sediments, resulting in severe bending of seismic ray paths) is used for the half-space beneath the top of salt. The reflections from the base of the salt body then appear, although the underlying sediments are very poorly imaged. Finally, the velocity model within these sediments is modified until an acceptable image is obtained.

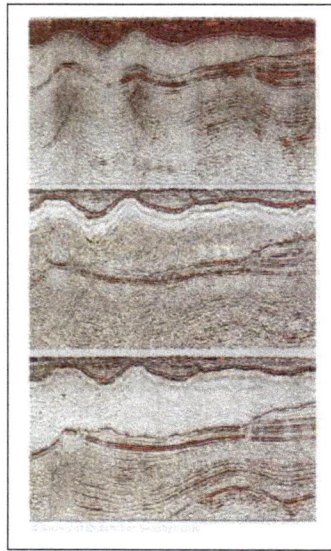

Improvements in imaging using different migration techniques. The upper part shows a result of imaging beneath salt using prestack time migration; the middle part uses post-stack depth migration; and the bottom uses prestack depth migration. Note the increasing ability to image sediments below the salt body.

Trace Inversion for Impedance

Seismic reflections at zero offset result from contrasts in acoustic impedance, involving just the P-wave velocity and density of the layers at the interface. If we can identify the seismic wavelet that propagated through the earth and reflected from the layer contrasts, we can then remove the effect of that wavelet and obtain a series of reflection coefficients at the interfaces. Then, we can simply integrate these reflection coefficients and determine the acoustic impedance in the layers between the interfaces. This "inversion" procedure leads us to a seismic volume that portrays layer properties (in terms of impedance) rather than interface characteristics, and assumes that the reflecting horizons have already been properly migrated to their appropriate positions.

Acoustic Impedance

If the seismic data were noise-free and contained all frequencies, from zero frequency (infinite wavelength) to very high frequencies (short wavelengths), the solution should be unique, but seismic data are noisy and band-limited and do not contain the very lowest frequencies nor the higher frequencies that are often of interest. A number of methods have been developed to overcome these shortcomings, including a "sparse-spike" inversion, in which the trade-off between the number of reflecting horizons and "noise" is chosen by the investigator (a technique that simultaneously solves for the "background" velocity trend and the impedance contrasts) and statistical or neural-network techniques that relate seismic features to properties inferred from borehole data. To a greater or lesser degree, these techniques rely on borehole sonic logs or on other velocity information or assumptions to incorporate long-wavelength velocity models (the background velocity trend). In general, a calibrated and competently processed inversion volume can be of considerable use to the interpreter or the engineer, providing insight to layer properties and continuity, which may not be apparent from the traditional reflection-seismic display; in particular, the thinner beds are usually more distinctly identified through removal of wavelet tuning (interference of reflections from the top and bottom of the bed) and subtle changes in impedance that are not easily recognized in the reflection image that can be seen in the inverted volume. Because the inversion process results in volume properties, rather than interface properties, it is possible to isolate and image individual bodies within certain impedance ranges. An example of the results of body-capture after a sparse-spike inversion, intended to identify hydrocarbon reservoirs, is shown in figure.

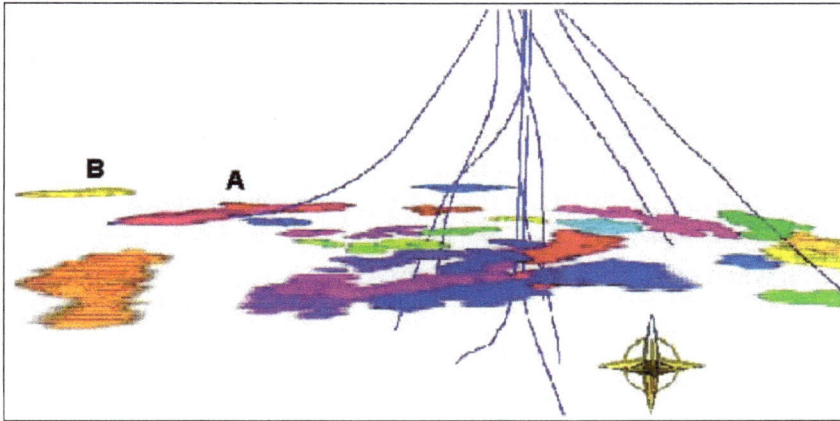

An example of sparse-spike inversion results. The results of trace inversion can be used to identify spatially distinct bodies with specific impedance ranges. This example is of the same area of the Gulf of Mexico as shown in figure, but after a sparse-spike inversion routine has been performed and the lowest-impedance areas selected as individual bodies. In this area, virtually all these bodies are hydrocarbon reservoirs, although not all are large enough to be economically produced. The two reservoirs identified by A and B in figure are also identified here.

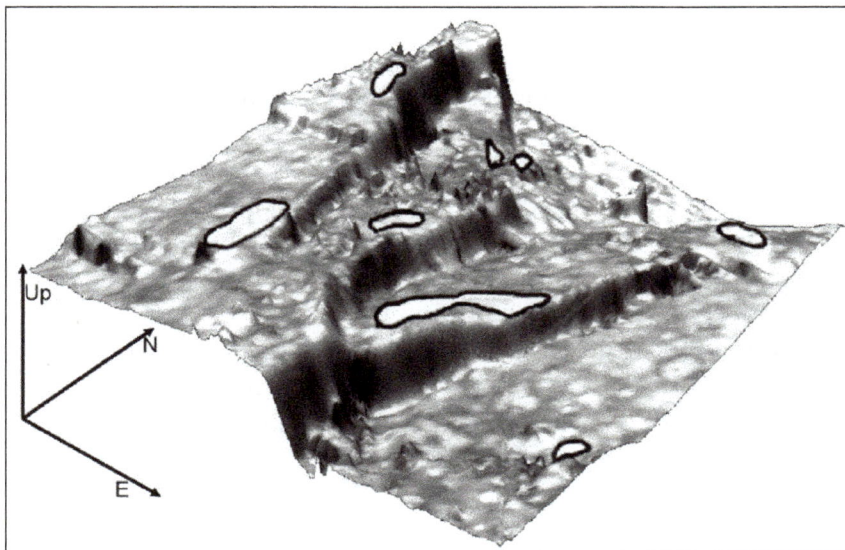

Amplitudes resulting from changes in seismic impedance. A perspective view of a single horizon containing several potential reservoirs is shown from the Teal South area of the Gulf of Mexico. The coloring is based on the amplitude of the reflected arrival at this horizon, with the hotter colors indicating larger (negative) amplitudes, resulting (in this case) from high-GOR oil in both produc-ing and unproduced reservoirs. The reservoirs have been highlighted for increased visibility on the black-and-white version of a typically color display.

In general, it is appropriate to invert only true zero-offset seismic data for acoustic impedance because the nonzero offsets are influenced by other parameters, notably the ratio between the P-wave velocity and the S-wave velocity (or, alternatively, Poisson's ratio). Yet typical seismic data has been processed by stacking all appropriate offsets after correcting for normal moveout and muting, and the amplitude of each reflection represents a sort of average amplitude over all of the offsets used. In many cases, this distinction is not important because the amplitude normally decays slightly with offset (after routine correction for geometric spreading) and affects all stacked samples similarly, but for many cases, and especially those of most interest, the amplitudes vary with offset. Inverting a seismic section containing stacked data does not always yield a true acous-tic impedance volume. (The term "acoustic" refers to compressional-wave effects only, and acous-tic models assume that the material does not propagate shear waves or that shear waves are not of any significance in wave transmission.) In practice, this is true for seismic compressional waves at normal incidence but is not valid for compressional waves at non-normal incidence in a solid material because of partial conversion to reflected and refracted shear waves. The term "elastic" is used to describe models incorporating compressional and shear effects. Thus, if we interpret a stacked seismic volume that has been inverted for acoustic impedance, we have implicitly assumed that the offsets used in stacking were small and/or that the offset-dependence of amplitudes is negligible. In the cases where these assumptions are not true, we must recognize that the values of acoustic impedance resulting from the inversion process are not precise; in fact, the disagreement of the acoustic inversion results, with a model based on well logs, is often an indication of AVO effects and can be used as an exploration tool.

Elastic Impedance

In order to separate the acoustic model (compressional-wave only) from the elastic model (in-cluding shear effects), the inversion process can be conducted on two or three different stacked seismic volumes, each composed of traces that resulted from stacking a different range of offsets. The volume created from traces in the near-offset range (or a volume made by extrapolating the AVO behavior to zero offset at each sample) is inverted to obtain the acoustic impedance volume. A volume created from traces in the far-offset range is inverted to obtain a new impedance volume called the "elastic impedance". The elastic impedance volume includes the effects of the compres-sional impedance and the AVO behavior resulting from the V_p/V_s ratio; the two volumes can be interpreted jointly to obtain the desired fluid or lithology indicator sought. Just as in AVO studies, one can also try to obtain a three-parameter inversion, using three different offset ranges and, for example, solve for compressional/shear velocities and density. Converted-wave data can also be inverted for shear impedance.

Nomenclature

V_p = P-wave velocity

V_s = S-wave velocity

Seismic Waves

Seismic waves are the resulting waves from earthquakes, volcanic eruptions, magma movement, large landslides and large man-made explosions. They can be categorized into surface waves, P-waves, S-waves, Stoneley waves, etc. The topics elaborated in this chapter will help in gaining a better perspective about these seismic waves.

Seismic wave is the vibration generated by an earthquake, explosion, or similar energetic source and propagated within the Earth or along its surface. Earthquakes generate four principal types of elastic waves; two, known as body waves, travel within the Earth, whereas the other two, called surface waves, travel along its surface. Seismographs record the amplitude and frequency of seismic waves and yield information about the Earth and its subsurface structure. Artificially generated seismic waves recorded during seismic surveys are used to collect data in oil and gas prospecting and engineering.

Of the body waves, the primary, or P, wave has the higher speed of propagation and so reaches a seismic recording station faster than the secondary, or S, wave. P waves, also called compressional or longitudinal waves, give the transmitting medium — whether liquid, solid, or gas — a back-and-forth motion in the direction of the path of propagation, thus stretching or compressing the medium as the wave passes any one point in a manner similar to that of sound waves in air. In the Earth, P waves travel at speeds from about 6 km (3.7 miles) per second in surface rock to about 10.4 km (6.5 miles) per second near the Earth's core some 2,900 km (1,800 miles) below the surface. As the waves enter the core, the velocity drops to about 8 km (5 miles) per second. It increases to about 11 km (6.8 miles) per second near the centre of the Earth. The speed increase with depth results from increased hydrostatic pressure as well as from changes in rock composition; in general, the increase causes P waves to travel in curved paths that are concave upward.

S waves, also called shear or transverse waves, cause points of solid media to move back and forth perpendicular to the direction of propagation; as the wave passes, the medium is sheared first in one direction and then in another. In the Earth the speed of S waves increases from about 3.4 km (2.1 miles) per second at the surface to 7.2 km (4.5 miles) per second near the boundary of the core, which, being liquid, cannot transmit them; indeed, their observed absence is a compelling argument for the liquid nature of the outer core. Like P waves, S waves travel in curved paths that are concave upward.

Of the two surface seismic waves, Love waves — named after the British seismologist A.E.H. Love, who first predicted their existence — travel faster. They are propagated when the solid medium near the surface has varying vertical elastic properties. Displacement of the medium by the wave is entirely perpendicular to the direction of propagation and has no vertical or longitudinal components. The energy of Love waves, like that of other surface waves, spreads from the source in two directions rather than in three, and so these waves produce a strong record at seismic stations even when originating from distant earthquakes.

The other principal surface waves are called Rayleigh waves after the British physicist Lord Rayleigh, who first mathematically demonstrated their existence. Rayleigh waves travel along the free surface of an elastic solid such as the Earth. Their motion is a combination of longitudinal compression and dilation that results in an elliptical motion of points on the surface. Of all seismic waves, Rayleigh waves spread out most in time, producing a long wave duration on seismographs.

SURFACE WAVES

Surface waves are typically generated when the source of the earthquake is close to the Earth's surface. As their name suggests, surface waves travel just below the surface of the ground. Although they move even more slowly than S-waves, they can be much larger in amplitude and are often the most destructive type of seismic wave. There are several types of surface wave, but the two most common varieties are Rayleigh waves and Love waves.

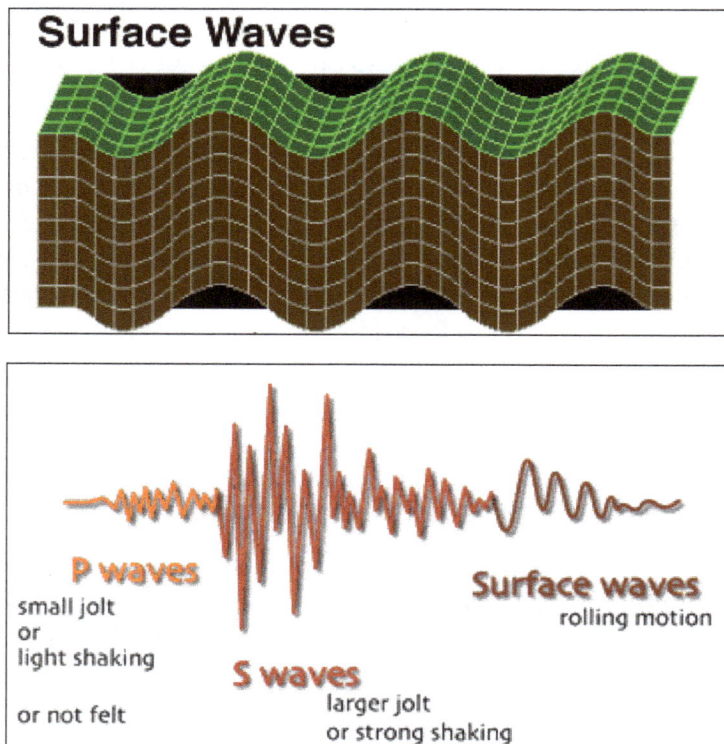

Surface Wave Methods

The assumption that the depth of investigation is equal to one-half of the wavelength can be used to generate a velocity profile with depth. This last assumption is somewhat supported by surface wave theory, but more modern and comprehensive methods are available for inversion of Rayleigh-wave observations. Similar data can be obtained from impulsive sources if the recording is made at sufficient distance such that the surface wave train has separated into its separate frequency components.

Spectral Analysis of Surface Waves (SASW)

The promise, both theoretical and observational, of surface wave methods has resulted in significant applications of technology to their exploitation. The problem is twofold:

- To determine, as a function of frequency, the velocity of surface waves traveling along the surface (this curve, often presented as wavelength versus phase velocity, is called a dispersion curve).

- From the dispersion curve, determine an earth structure that would exhibit such dispersion. This inversion, which is ordinarily done by forward modeling, has been automated with varying degrees of success.

Basic Concept

Spectral analysis, via the Fourier transform, can convert any time-domain function into its constituent frequencies. Cross-spectral analysis yields two valuable outputs from the simultaneous spectral analysis of two time functions. One output is the phase difference between the two time functions as a function of frequency. This phase difference spectrum can be converted to a time difference (as a function of frequency) by use of the relationship:

$$\Delta t(f) = \frac{\Phi(f)}{2\pi f},$$

where,

$\Delta t(f)$ = frequency-dependent time difference.

$\Phi(f)$ = cross-spectral phase at frequency f.

f = frequency to which the time difference applies.

If the two time functions analyzed are the seismic signals recorded at two geophones a distance d apart, then the velocity, as a function of frequency, is given by:

$$V(f) = \frac{d}{t(f)},$$

where,

d = distance between geophones.

t(f) = term determined from the cross-spectral phase.

If the wavelength (λ) is required, it is given by:

$$\lambda(f) = \frac{V(f)}{f},$$

As these mathematical operations are carried at for a variety of frequencies, an extensive dispersion curve is generated. The second output of the cross-spectral analysis that is useful in this work is the coherence function. This output measures the similarity of the two inputs as a function of frequency. Normalized to lie between 0 and 1, a coherency of greater than 0.9 is often required for effective phase difference estimates. Once the dispersion curve is in hand, the calculation-intensive inversion process can proceed. Although the assumption given above of depth equal to one-half the wavelength may be adequate if relatively few data are available, the direct calculation of a sample dispersion curve from a layered model is necessary to account for the abundance of data that can be recorded by a modern seismic system. Whether or not the inversion is automated, the requirements for a good geophysical inversion should be followed, and more observations than parameters should be selected.

Calculation methods for the inversion are beyond the scope of this manual. The model used is a set of flat-lying layers made up of thicknesses and shear-wave velocities. More layers are typically used than are suspected to be present, and one useful iteration is to consolidate the model layers into a geologically consistent model and repeat the inversion for the velocities only.

The advantages of this method are:

- High frequencies (1-300 Hz) can be used, resulting in definition of very thin layers.

- The refraction requirement of increasing velocities with depth is not present; thus, velocities that decrease with depth are detectable.

By using both of these advantages, this method has been used to investigate pavement substrate strength. An example of typical data obtained by an SASW experiment is shown in figure. The scatter of these data is smaller than typical SASW data. Models obtained by two different inversion schemes are shown in figure along with some crosshole data for comparison. Note that the agreement is excellent above 20 m of depth.

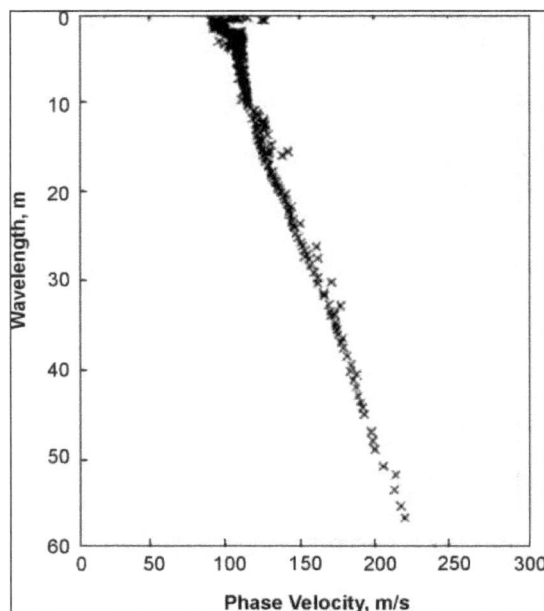

Typical spectral analysis of surface waves data.

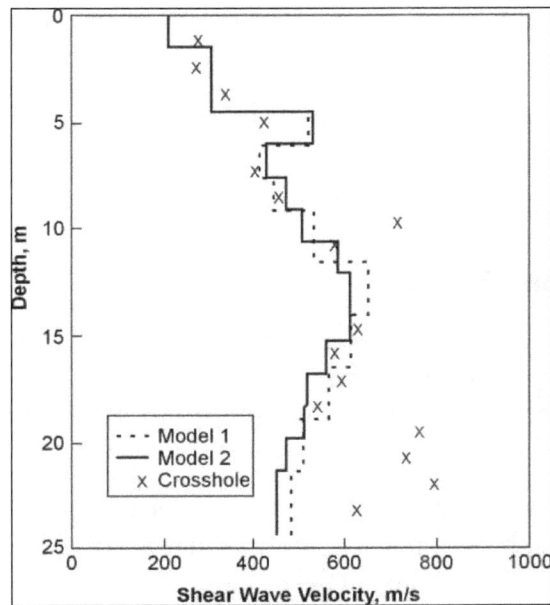

Inversion results of typical spectral analysis of surface waves data.

Data Acquisition

Most crews are equipped with a two- or four-channel spectrum analyzer, which provides the cross-spectral phase and coherence information. The degree of automation of the subsequent processing varies widely from laborious manual entry of the phase velocities into an analysis program to automated acquisition and preliminary processing. The inversion process similarly can be based on forward modeling with lots of human interaction or true inversion by computer after some manual smoothing.

A typical SASW crew consists of two persons, one to operate and coordinate the source and one to monitor the quality of the results. Typical field procedures are to place two (or four) geophones or accelerometers close together and to turn on the source. The source may be any mechanical source of high-frequency energy; moving bulldozers, dirt whackers, hammer blows, and vibrators have been used. Some discretion is advised as the source must operate for long periods of time, and the physics of what is happening are important. Rayleigh waves have predominantly vertical motion; thus, a source whose impedance is matched to the soil and whose energy is concentrated in the direction and frequency band of interest will be more successful.

Phase velocities are determined for waves with wavelength from 0.5 to 3 times the distance between the geophones. Then the phones are moved apart, usually increasing the separation by a factor of two. Thus, overlapping data are acquired, and the validity of the process is checked. This process continues until the wavelength being measured is equal to the required depth of investigation. Then the apparatus is moved to the next station where a sounding is required. After processing, a vertical profile of the shear-wave velocities is produced.

Advantages and Limitations

- The assumption of plane layers from the source to the recording point may not be accurate.

- Higher modes of the Rayleigh wave may be recorded. The usual processing assumption is that the fundamental mode has been measured.

- Spreading the geophones across a lateral inhomogeneity will produce complications beyond the scope of the method.

- Very high frequencies may be difficult to generate and record.

Common-offset Rayleigh Wave Method

This method is also called Common-offset Surface Waves. The method is quite effective at mapping inhomogeneities in the near surface, although it is not a frequently used method. The field techniques are easy to apply, and compared to other seismic methods, it is a rapid technique.

This method uses Rayleigh waves to detect fracture zones and associated voids. Rayleigh waves, also known as surface waves, have a particle motion that is counterclockwise with respect to the direction of travel. Figure illustrates the particle motion for Rayleigh waves traveling in the positive X direction. In addition, the particle displacement is greatest at the ground surface, near the Rayleigh wave source, and decreases with depth. Three shot points are shown, labeled A, B, and C. The particle motion and displacement are shown for five depths under each shot point. For shot B, over the void, no Rayleigh waves are transmitted through the water/air filled void. This affects the measured Rayleigh wave recorded by the geophones over the void. Four parameters are usually observed. The first is an increase in the travel time of the Rayleigh wave as the fracture zone above the void is crossed. The second parameter is a decrease in the amplitude of the Rayleigh wave. The third parameter is reverberations (sometimes called ringing) as the void is crossed. The fourth parameter is a shift in the peak frequency toward lower frequencies. This is caused by trapped waves, similar to a tube wave in a borehole. The effective depth of penetration is approximately one-third to one-half of the wavelength of the Rayleigh wave.

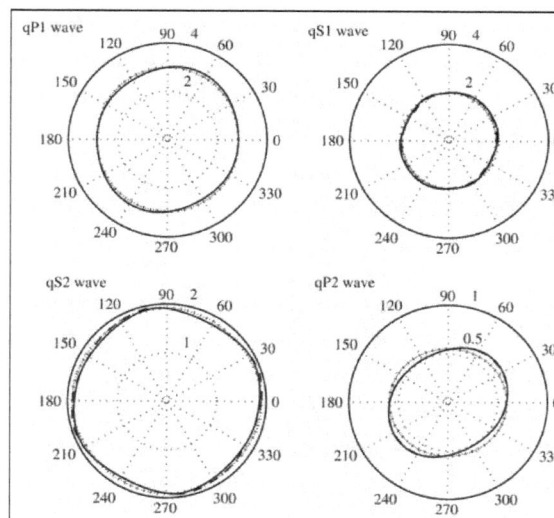

Rayleigh wave particle motion and displacement.

Rayleigh waves are created by any impact source. For shallow investigations, a hammer is all that is needed. Data are recorded using one geophone and one shot point. The distance between the shot and geophone depends on the depth of investigation and is usually about twice the expected target depth. Data are recorded at regular intervals across the traverse while maintaining the same shot geophone separation. The interval between stations depends on the expected size of the void/fracture zone and the desired resolution. Generally, in order to clearly see the void/fracture zone, it is desirable to have several stations that cross the area of interest. Figure presents data from a common offset Rayleigh wave survey over a void/fracture zone in an alluvial basin. The geophone traces are drawn horizontally with the vertical axis being distance (shot stations).

The data may be filtered to highlight the Rayleigh wave frequencies and is then plotted as shown in figure. Because the peak frequency of the seismic signal decreases over a void/fracture zone, a spectral analysis of the traces can assist in the interpretation of the data by highlighting the traces with lower frequencies. The data shown in figure illustrates many of the features expected over a void/fracture zone. The travel time to the first arrival of the Rayleigh wave is greater across the void/fracture and is wider than the actual fractured zone. The amplitudes of the Rayleigh waves decrease as the zone is crossed. In addition, the wavelength of the signals over the fracture/void is longer than those over unfractured rock, showing that the high frequencies have been attenuated. Since the records are not long enough, the ringing effect is not presented in these data.

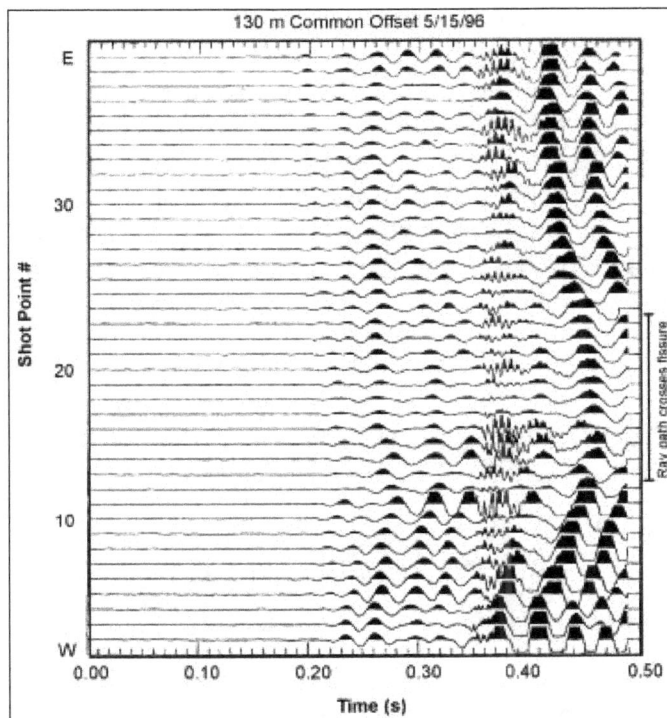

Data from a Rayleigh wave survey over a void/fracture zone.

Rayleigh waves are influenced by the shear strength of the rocks as well as fractures and voids, and changes in shear strength can cause anomalies similar to those obtained over these features. The depth of penetration and target resolution are influenced by the wavelengths generated by the seismic source. Since longer wavelengths, which have lower resolution, are needed to investigate to greater depths, fractures and voids at depth need to be increasingly larger in order to be observed. However, this method is faster and less costly than most other seismic methods.

Love Waves

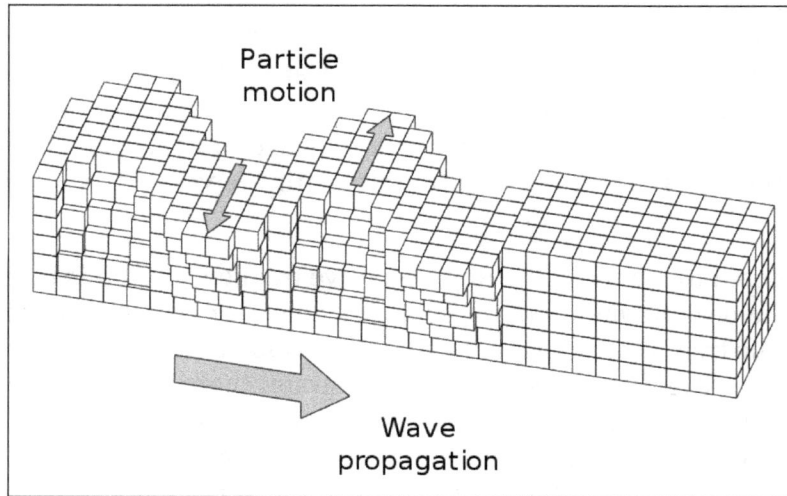

How Love waves work.

In elastodynamics, Love waves, named after Augustus Edward Hough Love, are horizontally polarized surface waves. The Love wave is a result of the interference of many shear waves (S–waves) guided by an elastic layer, which is welded to an elastic half space on one side while bordering a vacuum on the other side. In seismology, Love waves (also known as Q waves) are surface seismic waves that cause horizontal shifting of the Earth during an earthquake. Augustus Edward Hough Love predicted the existence of Love waves mathematically in 1911. They form a distinct class, different from other types of seismic waves, such as P-waves and S-waves (both body waves), or Rayleigh waves (another type of surface wave). Love waves travel with a lower velocity than P- or S-waves, but faster than Rayleigh waves. These waves are observed only when there is a low velocity layer overlying a high velocity layer/sub-layers.

The particle motion of a Love wave forms a horizontal line perpendicular to the direction of propagation (i.e. are transverse waves). Moving deeper into the material, motion can decrease to a "node" and then alternately increase and decrease as one examines deeper layers of particles. The amplitude, or maximum particle motion, often decreases rapidly with depth.

Since Love waves travel on the Earth's surface, the strength (or amplitude) of the waves decrease exponentially with the depth of an earthquake. However, given their confinement to the surface, their amplitude decays only as $\frac{1}{\sqrt{r}}$, where r represents the distance the wave has travelled from the earthquake. Surface waves therefore decay more slowly with distance than do body waves, which travel in three dimensions. Large earthquakes may generate Love waves that travel around the Earth several times before dissipating.

Since they decay so slowly, Love waves are the most destructive outside the immediate area of the focus or epicentre of an earthquake. They are what most people feel directly during an earthquake.

In the past, it was often thought that animals like cats and dogs could predict an earthquake before it happened. However, they are simply more sensitive to ground vibrations than humans and able to detect the subtler body waves that precede Love waves, like the P-waves and the S-waves.

Basic Theory

The conservation of linear momentum of a linear elastic material can be written as:

$$\nabla \cdot (\mathbf{C} : \nabla \mathbf{u}) = \rho \ddot{\mathbf{u}}$$

where $\mathbf{u}$ is the displacement vector and $\mathbf{C}$ is the stiffness tensor. Love waves are a special solution ($\mathbf{u}$) that satisfy this system of equations. We typically use a Cartesian coordinate system (x, y, z) to describe Love waves.

Consider an isotropic linear elastic medium in which the elastic properties are functions of only the z coordinate, i.e. the Lamé parameters and the mass density can be expressed as $\lambda(z), \mu(z), \rho(z)$. Displacements (u, v, w) produced by Love waves as a function of time (t) have the form:

$$u(x, y, z, t) = 0 \ , \ v(x, y, z, t) = \hat{v}(x, z, t) \ , \ w(x, y, z, t) = 0.$$

These are therefore antiplane shear waves perpendicular to the (x, z) plane. The function $\hat{v}(x, z, t)$ can be expressed as the superposition of harmonic waves with varying wave numbers (k) and frequencies (ω). Consider a single harmonic wave:

$$\hat{v}(x, z, t) = V(k, z, \omega) \ \exp[i(kx - \omega t)]$$

where $i = \sqrt{-1}$. The stresses caused by these displacements are:

$$\sigma_{xx} = 0, \ \ \sigma_{yy} = 0, \ \ \sigma_{zz} = 0, \ \ \tau_{zx} = 0, \ \ \tau_{yz} = \mu(z) \frac{dV}{dz} \exp[i(kx - \omega t)], \ \ \tau_{xy} = ik\mu(z)V(k, z, \omega) \exp[i(kx - \omega t)].$$

If we substitute the assumed displacements into the equations for the conservation of momentum, we get a simplified equation:

$$\frac{d}{dz}\left[\mu(z) \frac{dV}{dz} \right] = [k^2 \mu(z) - \omega^2 \rho(z)]V(k, z, \omega).$$

The boundary conditions for a Love wave are that the surface tractions at the free surface ($z = 0$) must be zero. Another requirement is that the stress component τ_{yz} in a layer medium must be continuous at the interfaces of the layers. To convert the second order differential equation in V into two first order equations, we express this stress component in the form:

$$\tau_{yz} = T(k, z, \omega)\exp[i(kx - \omega t)]$$

to get the first order conservation of momentum equations:

$$\frac{d}{dz}\begin{bmatrix} V \\ T \end{bmatrix} = \begin{bmatrix} 0 & 1/\mu(z) \\ k^2 \mu(z) - \omega^2 \rho(z) & 0 \end{bmatrix}\begin{bmatrix} V \\ T \end{bmatrix}.$$

The above equations describe an eigenvalue problem whose solution eigenfunctions can be found by a number of numerical methods. Another common, and powerful, approach is the propagator matrix method (also called the matricant approach).

Rayleigh Wave

Rayleigh waves are a type of surface acoustic wave that travel along the surface of solids. They can be produced in materials in many ways, such as by a localized impact or by piezo-electric transduction, and are frequently used in non-destructive testing for detecting defects. Rayleigh waves are part of the seismic waves that are produced on the Earth by earthquakes. When guided in layers they are referred to as Lamb waves, Rayleigh-Lamb waves, or generalized Rayleigh waves.

Characteristics

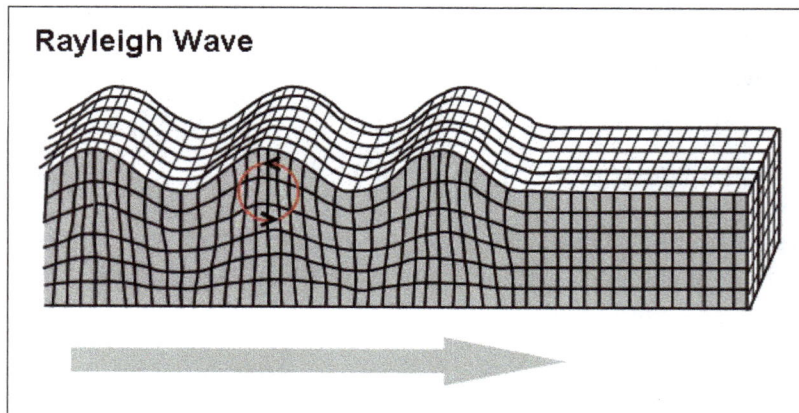

Picture of a Rayleigh wave.

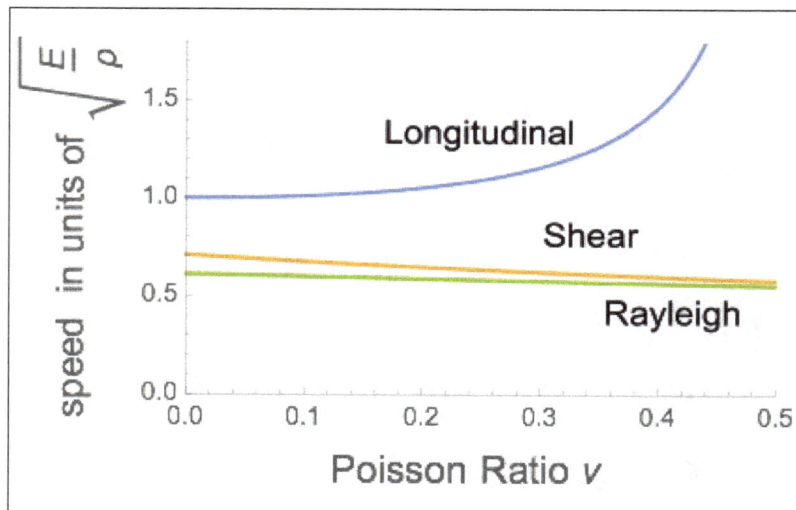

Comparison of the Rayleigh wave speed with shear and longitudinal wave speeds for an isotropic elastic material. The speeds are shown in dimensionless units.

Rayleigh waves are a type of surface wave that travel near the surface of solids. Rayleigh waves include both longitudinal and transverse motions that decrease exponentially in amplitude as distance from the surface increases. There is a phase difference between these component motions.

The existence of Rayleigh waves was predicted in 1885 by Lord Rayleigh, after whom they were named. In isotropic solids these waves cause the surface particles to move in ellipses in planes normal to the surface and parallel to the direction of propagation – the major axis of the ellipse is vertical. At the surface and at shallow depths this motion is *retrograde*, that is the in-plane motion of a particle is counterclockwise when the wave travels from left to right. At greater depths the particle motion becomes *prograde*. In addition, the motion amplitude decays and the eccentricity changes as the depth into the material increases. The depth of significant displacement in the solid is approximately equal to the acoustic wavelength. Rayleigh waves are distinct from other types of surface or guided acoustic waves such as Love waves or Lamb waves, both being types of guided waves supported by a layer, or longitudinal and shear waves, that travel in the bulk.

Rayleigh waves have a speed slightly less than shear waves by a factor dependent on the elastic constants of the material. The typical speed of Rayleigh waves in metals is of the order of 2–5 km/s, and the typical Rayleigh speed in the ground is of the order of 50–300 m/s. For linear elastic materials with positive Poisson ratio ($v > 0$), the Rayleigh wave speed can be approximated as $c_R / c_S = \dfrac{0.862 + 1.14v}{1+v}$. Since Rayleigh waves are confined near the surface, their in-plane amplitude when generated by a point source decays only as $1/\sqrt{r}$, where r is the radial distance. Surface waves therefore decay more slowly with distance than do bulk waves, which spread out in three dimensions from a point source. This slow decay is one reason why they are of particular interest to seismologists. Rayleigh waves can circle the globe multiple times after a large earthquake and still be measurably large. There is a difference in the behavior (Rayleigh wave velocity, displacements, trajectories of the particle motion, stresses) of Rayleigh surface waves with positive and negative Poisson's ratio.

In seismology, Rayleigh waves (called "ground roll") are the most important type of surface wave, and can be produced (apart from earthquakes), for example, by ocean waves, by explosions, by railway trains and ground vehicles, or by a sledgehammer impact.

Rayleigh Wave Dispersion

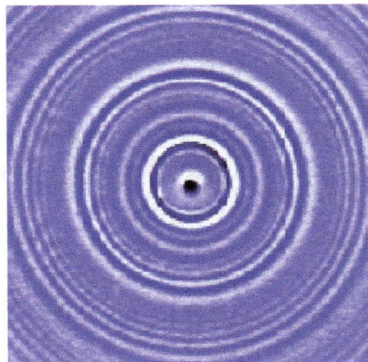

Dispersion of Rayleigh waves in a thin gold film on glass.

In isotropic, linear elastic materials described by Lamé coefficients   and μ, Rayleigh waves have a speed given by solutions to the equation:

$$\zeta^3 - 8\zeta^2 + 8\zeta(3 - 2\eta) - 16(1 - \eta) = 0,$$

where $\zeta = \omega^2 / k^2 \beta^2$, $\eta = \beta^2 / \alpha^2$, $\rho\alpha^2 = \lambda + 2\mu$ and $\rho\beta^2 = \mu$. Since this equation has no inherent scale, the boundary value problem giving rise to Rayleigh waves are dispersionless. An interesting special case is the Poisson solid, for which $\lambda = \mu$, since this gives a frequency-independent phase velocity equal to $\omega/k = \beta\sqrt{0.8453}$.

The elastic constants often change with depth, due to the changing properties of the material. This means that the velocity of a Rayleigh wave in practice becomes dependent on the wavelength (and therefore frequency), a phenomenon referred to as dispersion. Waves affected by dispersion have a different wave train shape. Rayleigh waves on ideal, homogeneous and flat elastic solids show no dispersion, as stated above. However, if a solid or structure has a density or sound velocity that varies with depth, Rayleigh waves become dispersive. One example is Rayleigh waves on the Earth's surface: those waves with a higher frequency travel more slowly than those with a lower frequency. This occurs because a Rayleigh wave of lower frequency has a relatively long wavelength. The displacement of long wavelength waves penetrates more deeply into the Earth than short wavelength waves. Since the speed of waves in the Earth increases with increasing depth, the longer wavelength (low frequency) waves can travel faster than the shorter wavelength (high frequency) waves. Rayleigh waves thus often appear spread out on seismograms recorded at distant earthquake recording stations. It is also possible to observe Rayleigh wave dispersion in thin films or multi-layered structures.

Rayleigh Waves in Non-destructive Testing

Rayleigh waves are widely used for materials characterization, to discover the mechanical and structural properties of the object being tested – like the presence of cracking, and the related shear modulus. This is in common with other types of surface waves. The Rayleigh waves used for this purpose are in the ultrasonic frequency range.

They are used at different length scales because they are easily generated and detected on the free surface of solid objects. Since they are confined in the vicinity of the free surface within a depth (wavelength) linked to the frequency of the wave, different frequencies can be used for characterization at different length scales.

Rayleigh Waves in Electronic Devices

Rayleigh waves propagating at high ultrasonic frequencies (10–1000 MHz) are used widely in different electronic devices. In addition to Rayleigh waves, some other types of surface acoustic waves (SAW), e.g. Love waves, are also used for this purpose. Examples of electronic devices using Rayleigh waves are filters, resonators, oscillators, sensors of pressure, temperature, humidity, etc. Operation of SAW devices is based on the transformation of the initial electric signal into a surface wave that, after achieving the required changes to the spectrum of the initial electric signal as a result of its interaction with different types of surface inhomogeneity, is transformed back into a modified electric signal. The transformation of the initial electric energy into mechanical energy (in the form of SAW) and back is usually accomplished via the use of piezoelectric materials for both generation and reception of Rayleigh waves as well as for their propagation.

Rayleigh Waves in Geophysics

Rayleigh Waves from Earthquakes

Because Rayleigh waves are surface waves, the amplitude of such waves generated by an earthquake generally decreases exponentially with the depth of the hypocenter (focus). However, large earthquakes may generate Rayleigh waves that travel around the Earth several times before dissipating.

In seismology longitudinal and shear waves are known as P-waves and S-waves, respectively, and are termed body waves. Rayleigh waves are generated by the interaction of P- and S-waves at the surface of the earth, and travel with a velocity that is lower than the P-, S-, and Love wave velocities. Rayleigh waves emanating outward from the epicenter of an earthquake travel along the surface of the earth at about 10 times the speed of sound in air (0.340 km/s), that is 3 km/s.

Due to their higher speed, the P- and S-waves generated by an earthquake arrive before the surface waves. However, the particle motion of surface waves is larger than that of body waves, so the surface waves tend to cause more damage. In the case of Rayleigh waves, the motion is of a rolling nature, similar to an ocean surface wave. The intensity of Rayleigh wave shaking at a particular location is dependent on several factors:

- The size of the earthquake.
- The distance to the earthquake.
- The depth of the earthquake.
- The geologic structure of the crust.
- The focal mechanism of the earthquake.
- The rupture directivity of the earthquake.

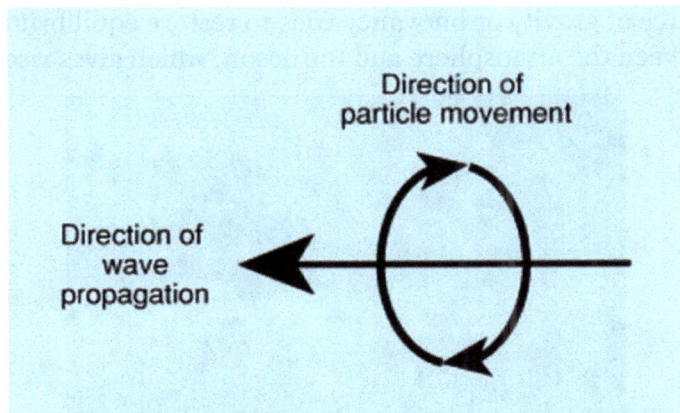

Rayleigh wave direction.

Local geologic structure can serve to focus or defocus Rayleigh waves, leading to significant differences in shaking over short distances.

Rayleigh Waves in Seismology

Low frequency Rayleigh waves generated during earthquakes are used in seismology to characterise the Earth's interior. In intermediate ranges, Rayleigh waves are used in geophysics and geotechnical engineering for the characterisation of oil deposits. These applications are based on the geometric dispersion of Rayleigh waves and on the solution of an inverse problem on the basis of seismic data collected on the ground surface using active sources (falling weights, hammers or small explosions, for example) or by recording microtremors. Rayleigh ground waves are important also for environmental noise and vibration control since they make a major contribution to traffic-induced ground vibrations and the associated structure-borne noise in buildings.

Other Manifestations

Animals

Low frequency (< 20 Hz) Rayleigh waves are inaudible, yet they can be detected by many mammals, birds, insects and spiders. Humans should be able to detect such Rayleigh waves through their Pacinian corpuscles, which are in the joints, although people do not seem to consciously respond to the signals. Some animals seem to use Rayleigh waves to communicate. In particular, some biologists theorize that elephants may use vocalizations to generate Rayleigh waves. Since Rayleigh waves decay slowly, they should be detectable over long distances. Note that these Rayleigh waves have a much higher frequency than Rayleigh waves generated by earthquakes.

After the 2004 Indian Ocean earthquake, some people have speculated that Rayleigh waves served as a warning to animals to seek higher ground, allowing them to escape the more slowly traveling tsunami. At this time, evidence for this is mostly anecdotal. Other animal early warning systems may rely on an ability to sense infrasonic waves traveling through the air.

Gravity Wave

In fluid dynamics, gravity waves are waves generated in a fluid medium or at the interface between two media when the force of gravity or buoyancy tries to restore equilibrium. An example of such an interface is that between the atmosphere and the ocean, which gives rise to wind waves.

Surface gravity wave, breaking on an ocean beach.

A gravity wave results when fluid is displaced from a position of equilibrium. The restoration of the fluid to equilibrium will produce a movement of the fluid back and forth, called a wave orbit. Gravity waves on an air–sea interface of the ocean are called surface gravity waves or surface waves, while gravity waves that are within the body of the water (such as between parts of different densities) are called internal waves. Wind-generated waves on the water surface are examples of gravity waves, as are tsunamis and ocean tides.

Wind-generated gravity waves on the free surface of the Earth's ponds, lakes, seas and oceans have a period of between 0.3 and 30 seconds (3Hz to 30mHz). Shorter waves are also affected by surface tension and are called gravity–capillary waves and (if hardly influenced by gravity) capillary waves. Alternatively, so-called infragravity waves, which are due to subharmonic nonlinear wave interaction with the wind waves, have periods longer than the accompanying wind-generated waves.

Wave clouds.

Atmospheric gravity waves seen from space.

Atmosphere Dynamics on Earth

In the Earth's atmosphere, gravity waves are a mechanism that produce the transfer of momentum from the troposphere to the stratosphere and mesosphere. Gravity waves are generated in the troposphere by frontal systems or by airflow over mountains. At first, waves propagate through the atmosphere without appreciable change in mean velocity. But as the waves reach more rarefied (thin) air at higher altitudes, their amplitude increases, and nonlinear effects cause the waves to break, transferring their momentum to the mean flow. This transfer of momentum is responsible for the forcing of the many large-scale dynamical features of the atmosphere. For example, this momentum transfer is partly responsible for the driving of the Quasi-Biennial Oscillation, and in the mesosphere, it is thought to be the major driving force of the Semi-Annual Oscillation. Thus, this process plays a key role in the dynamics of the middle atmosphere.

The effect of gravity waves in clouds can look like altostratus undulatus clouds, and are sometimes confused with them, but the formation mechanism is different.

Quantitative Description

Deep Water

The phase velocity c of a linear gravity wave with wavenumber is given by the formula:

$$c = \sqrt{\frac{g}{k}},$$

where g is the acceleration due to gravity. When surface tension is important, this is modified to:

$$c = \sqrt{\frac{g}{k} + \frac{\sigma k}{\rho}},$$

where σ is the surface tension coefficient and ρ is the density.

Phase-speed Derivation

The gravity wave represents a perturbation around a stationary state, in which there is no velocity. Thus, the perturbation introduced to the system is described by a velocity field of infinitesimally small amplitude, $(u'(x,z,t), w'(x,z,t))$. Because the fluid is assumed incompressible, this velocity field has the streamfunction representation:

$$\mathbf{u}' = (u'(x,z,t), w'(x,z,t)) = (\psi_z, -\psi_x),$$

where the subscripts indicate partial derivatives. In this derivation it suffices to work in two dimensions (x,z) where gravity points in the negative z-direction. Next, in an initially stationary incompressible fluid, there is no vorticity, and the fluid stays irrotational, hence $\nabla \times \mathbf{u}' = 0$. In the streamfunction representation, $\nabla^2 \psi = 0$. Next, because of the translational invariance of the system in the x-direction, it is possible to make the ansatz:

$$\psi(x,z,t) = e^{ik(x-ct)} \Psi(z),$$

where k is a spatial wavenumber. Thus, the problem reduces to solving the equation:

$$\left(D^2 - k^2\right)\Psi = 0, \quad D = \frac{d}{dz}.$$

We work in a sea of infinite depth, so the boundary condition is at $z = -\infty$. The undisturbed surface is at $z = 0$, and the disturbed or wavy surface is at $z = \eta$, where η is small in magnitude. If no fluid is to leak out of the bottom, we must have the condition:

$$u = D\Psi = 0, \quad \text{on } z = -\infty.$$

Hence, $\Psi = Ae^{kz}$ on $z \in (-\infty, \eta)$, where A and the wave speed c are constants to be determined from conditions at the interface.

Free-surface Condition

At the free surface $z = \eta(x,t)$, the kinematic condition holds:

$$\frac{\partial \eta}{\partial t} + u' \frac{\partial \eta}{\partial x} = w'(\eta).$$

Linearizing, this is simply:

$$\frac{\partial \eta}{\partial t} = w'(0),$$

where the velocity $w'(\eta)$ is linearized on to the surface $z = 0$. Using the normal-mode and stream-function representations, this condition is $c\eta = \Psi$, the second interfacial condition.

Pressure Relation across the Interface

For the case with surface tension, the pressure difference over the interface at $z = \eta$ is given by the Young–Laplace equation:

$$p(z = \eta) = -\sigma\kappa,$$

where σ is the surface tension and κ is the curvature of the interface, which in a linear approximation is:

$$\kappa = \nabla^2 \eta = \eta_{xx}.$$

Thus,

$$p(z = \eta) = -\sigma\eta_{xx}.$$

However, this condition refers to the total pressure (base+perturbed), thus:

$$\left[P(\eta) + p'(0) \right] = -\sigma\eta_{xx}.$$

(As usual, the perturbed quantities can be linearized onto the surface z=0.) Using hydrostatic balance, in the form:

$$P = -\rho gz + \text{Const.},$$

this becomes:

$$p = g\eta\rho - \sigma\eta_{xx}, \ \ \text{on } z = 0.$$

The perturbed pressures are evaluated in terms of stream functions, using the horizontal momentum equation of the linearized Euler equations for the perturbations,

$$\frac{\partial u'}{\partial t} = -\frac{1}{\rho}\frac{\partial p'}{\partial x}$$

to yield $p' = \rho c D\Psi$.

Putting this last equation and the jump condition together,

$$c\rho D\Psi = g\eta\rho - \sigma\eta_{xx}.$$

Substituting the second interfacial condition $c\eta = \Psi$ and using the normal-mode representation, this relation becomes:

$$c^2\rho D\Psi = g\Psi\rho + \sigma k^2\Psi.$$

Using the solution $\Psi = e^{kz}$, this gives:

$$c = \sqrt{\frac{g}{k} + \frac{\sigma k}{\rho}}.$$

Since $c = \omega / k$ is the phase speed in terms of the angular frequency ω and the wavenumber, the gravity wave angular frequency can be expressed as:

$$\omega = \sqrt{gk}.$$

The group velocity of a wave (that is, the speed at which a wave packet travels) is given by:

$$c_g = \frac{d\omega}{dk},$$

and thus for a gravity wave,

$$c_g = \frac{1}{2}\sqrt{\frac{g}{k}} = \frac{1}{2}c.$$

The group velocity is one half the phase velocity. A wave in which the group and phase velocities differ is called dispersive.

Shallow Water

Gravity waves traveling in shallow water (where the depth is much less than the wavelength), are non-dispersive: The phase and group velocities are identical and independent of wavelength and frequency. When the water depth is h,

$$c_p = c_g = \sqrt{gh}.$$

Generation of Ocean Waves by Wind

Wind waves, as their name suggests, are generated by wind transferring energy from the atmosphere to the ocean's surface, and capillary-gravity waves play an essential role in this effect. There are two distinct mechanisms involved, called after their proponents, Phillips and Miles.

In the work of Phillips, the ocean surface is imagined to be initially flat (*glassy*), and a turbulent wind blows over the surface. When a flow is turbulent, one observes a randomly fluctuating velocity field superimposed on a mean flow (contrast with a laminar flow, in which the fluid motion is ordered and smooth). The fluctuating velocity field gives rise to fluctuating stresses (both tangential and normal) that act on the air-water interface. The normal stress, or fluctuating pressure acts as a forcing term (much like pushing a swing introduces a forcing term). If the frequency and wavenumber (w,k) of this forcing term match a mode of vibration of the capillary-gravity wave (as derived above), then there is a resonance, and the wave grows in amplitude. As with other resonance effects, the amplitude of this wave grows linearly with time.

The air-water interface is now endowed with a surface roughness due to the capillary-gravity waves, and a second phase of wave growth takes place. A wave established on the surface either spontaneously as described above, or in laboratory conditions, interacts with the turbulent mean flow in a manner described by Miles. This is the so-called critical-layer mechanism. A critical layer forms at a height where the wave speed c equals the mean turbulent flow U. As the flow is turbulent, its mean profile is logarithmic, and its second derivative is thus negative. This is precisely the condition for the mean flow to impart its energy to the interface through the critical layer. This supply of energy to the interface is destabilizing and causes the amplitude of the wave on the interface to grow in time. As in other examples of linear instability, the growth rate of the disturbance in this phase is exponential in time.

This Miles-Phillips Mechanism process can continue until an equilibrium is reached, or until the wind stops transferring energy to the waves (i.e. blowing them along) or when they run out of ocean distance, also known as fetch length.

P-WAVE

A P-wave is one of the two main types of elastic body waves, called seismic waves in seismology. P-waves travel faster than other seismic waves and hence are the first signal from an earthquake to arrive at any affected location or at a seismograph. P-waves may be transmitted through gases, liquids, or solids.

Nomenclature

The name P-wave can stand for either pressure wave (as it is formed from alternating compressions and rarefactions) or primary wave (as it has high velocity and is therefore the first wave to be recorded by a seismograph).

Seismic Waves in the Earth

Primary and secondary waves are body waves that travel within the Earth. The motion and behavior of both *P*-type and *S*-type in the Earth are monitored to probe the interior structure of the Earth. Discontinuities in velocity as a function of depth are indicative of changes in phase or composition. Differences in arrival times of waves originating in a seismic event like an earthquake as a result of waves taking different paths allow mapping of the Earth's inner structure.

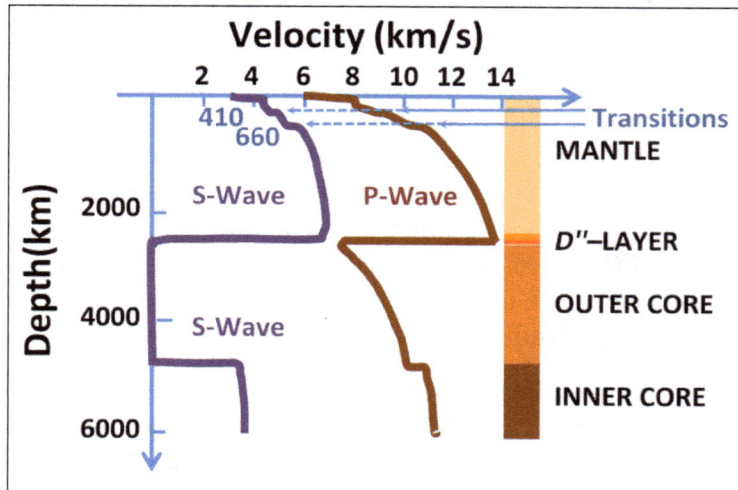

Velocity of seismic waves in the Earth versus depth. The negligible *S*-wave velocity in the outer core occurs because it is liquid, while in the solid inner core the *S*-wave velocity is non-zero.

P-wave Shadow Zone

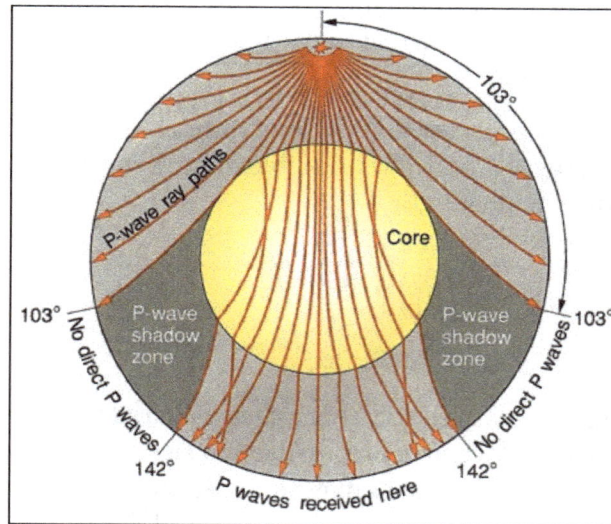

P-wave shadow zone.

Almost all the information available on the structure of the Earth's deep interior is derived from observations of the travel times, reflections, refractions and phase transitions of seismic body waves, or normal modes. P-waves travel through the fluid layers of the Earth's interior, and yet they are refracted slightly when they pass through the transition between the semisolid mantle and the liquid outer core. As a result, there is a P-wave "shadow zone" between 103° and 142° from the earthquake's focus, where the initial P-waves are not registered on seismometers. In contrast, S-waves do not travel through liquids.

As an Earthquake Warning

Advance earthquake warning is possible by detecting the nondestructive primary waves that travel more quickly through the Earth's crust than do the destructive secondary and Rayleigh waves.

The amount of advance warning depends on the delay between the arrival of the P-wave and other destructive waves, generally on the order of seconds up to about 60 to 90 seconds for deep, distant, large quakes such as the 2011 Tohoku earthquake. The effectiveness of advance warning depends on accurate detection of the P-waves and rejection of ground vibrations caused by local activity (such as trucks or construction). Earthquake early warning systems can be automated to allow for immediate safety actions, such as issuing alerts, stopping elevators at the nearest floors and switching off utilities.

Propagation

Velocity

In isotropic and homogeneous solids, a P-wave travels in a straight line longitudinal; thus, the particles in the solid vibrate along the axis of propagation (the direction of motion) of the wave energy. The velocity of P-waves in such a medium is given by:

$$v_p = \sqrt{\frac{K + \frac{4}{3}\mu}{\rho}} = \sqrt{\frac{\lambda + 2\mu}{\rho}}$$

where K is the bulk modulus (the modulus of incompressibility), μ is the shear modulus (modulus of rigidity, sometimes denoted as G and also called the second Lamé parameter), ρ is the density of the material through which the wave propagates, and λ is the first Lamé parameter.

In typical situations the interior of the Earth, the density ρ usually varies much less than K or μ, so the velocity is mostly "controlled" by these two parameters.

The elastic moduli P-wave modulus, M, is defined so that $M = K + 4\mu/3$ and thereby:

$$v_p = \sqrt{M/\rho} \ .$$

Typical values for P-wave velocity in earthquakes are in the range 5 to 8 km/s. The precise speed varies according to the region of the Earth's interior, from less than 6 km/s in the Earth's crust to 13.5 km/s in the lower mantle, and 11 km/s through the inner core.

Table: Velocity of Common Rock Types.

Rock Type	Velocity [m/s]	Velocity [ft/s]
Unconsolidated Sandstone	4600 - 5200	15000 - 17000
Consolidated Sandstone	5800	19000
Shale	1800 - 4900	6000 -16000
Limestone	5800 - 6400	19000 - 21000

Dolomite	6400 - 7300	21000 - 24000
Anhydrite	6100	20000
Granite	5800 - 6100	19000 - 20000
Gabbro	7200	23600

Geologist Francis Birch discovered a relationship between the velocity of P waves and the density of the material the waves are traveling in:

$$V_p = a(\bar{M}) + b\rho$$

which later became known as Birch's law.

S-WAVE

S-waves, secondary waves, or shear waves (sometimes called an elastic S-wave) are a type of elastic wave, and are one of the two main types of elastic body waves, so named because they move through the body of an object, unlike surface waves.

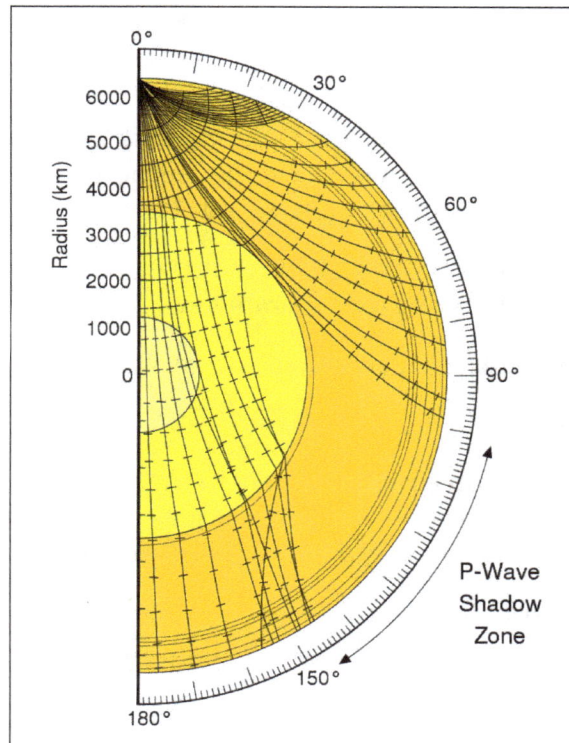

The shadow zone of a P-wave. S-waves don't penetrate the outer core, so they're shadowed everywhere more than 104° away from the epicentre.

The S-wave is a transverse wave, meaning that, in the simplest situation, the oscillations of the particles of the medium is perpendicular to the direction of wave propagation, and the main restoring force comes from shear stress. Therefore, S-waves can't propagate in liquids with zero (or very low) viscosity. However, it may propagate in liquids with high viscocity.

Its name, S for secondary, comes from the fact that it is the second direct arrival on an earthquake seismogram, after the compressional primary wave, or P-wave, because S-waves travel slower in rock. Unlike the P-wave, the S-wave cannot travel through the molten outer core of the Earth, and this causes a shadow zone for S-waves opposite to where they originate. They can still appear in the solid inner core: when a P-wave strikes the boundary of molten and solid cores at an oblique angle, S-waves will form and propagate in the solid medium. When these S-waves hit the boundary again at an oblique angle they will in turn create P-waves that propagate through the liquid medium. This property allows seismologists to determine some physical properties of the inner core.

In 1830, the mathematician Siméon Denis Poisson presented to the French Academy of Sciences an essay ("memoir") with a theory of the propagation of elastic waves in solids. In his memoir, he states that an earthquake would produce two different waves: one having a certain speed a and the other having a speed $a/\sqrt{3}$. At a sufficient distance from the source, when they can be considered plane waves in the region of interest, the first kind consists of expansions and compressions in the direction perpendicular to the wavefront (that is, parallel to the wave's direction of motion); while the second consists of stretching motions occurring in directions parallel to the front (perpendicular to the direction of motion).

Theory

Isotropic Medium

For the purpose of this explanation, a solid medium is considered isotropic if its strain (deformation) in response to stress is the same in all directions. Let $u = (u_1, u_2, u_3)$ be the displacement vector of a particle of such a medium from its "resting" position $x = (x_1, x_2, x_3)$ due elastic vibrations, understood to be a function of the rest position x and time t. The deformation of the medium at that point can be described the strain tensor e, the 3×3 matrix whose elements are:

$$e_{ij} = \frac{1}{2}(\partial_i u_j + \partial_j u_i)$$

where ∂_i denotes partial derivative with respect to position coordinate x_i. The strain tensor is related to the 3×3 stress tensor τ by the equation:

$$\tau_{ij} = \lambda \delta_{ij} \sum_k e_{kk} + 2\mu e_{ij}$$

Here δ_{ij} is the Kronecker delta (1 if $i = j$, 0 otherwise) and λ and μ are the Lamé parameters (μ being the material's shear modulus). It follows that:

$$\tau_{ij} = \lambda \delta_{ij} (\sum_k \partial_k u_k) + \mu(\partial_i u_j + \partial_j u_i)$$

From Newton's law of inertia, one also gets:

$$\rho \partial_t^2 u_i = \sum_j \partial_j \tau_{ij}$$

where ρ is the density (mass per unit volume) of the medium at that point, and ∂_t denotes partial derivative with respect to time. Combining the last two equations one gets the seismic wave equation in homogeneous media:

$$\rho \partial_t^2 u_i = \lambda \partial_i \left(\sum_k \partial_k u_k \right) + \mu \left(\sum_j \partial_i \partial_j u_j + \mu \, \partial_j \partial_j u_i \right).$$

Using the nabla operator notation of vector calculus, $\nabla = (\partial_1, \partial_2, \partial_3)$, with some approximations, this equation can be written as:

$$\rho \partial_t^2 \boldsymbol{u} = (\lambda + 2\mu) \nabla (\nabla \cdot \boldsymbol{u}) - \mu \nabla \times (\nabla \times \boldsymbol{u})$$

Taking the curl of this equation and applying vector identities, one gets:

$$\partial_t^2 (\nabla \times \mathbf{u}) = \frac{\mu}{\rho} \nabla^2 (\nabla \times \mathbf{u})$$

This formula is the wave equation applied to the vector quantity $\nabla \times \boldsymbol{u}$, which is the material's shear strain. Its solutions, the S-waves, are linear combinations of sinusoidal plane waves of various wavelengths and directions of propagation, but all with the same speed $\beta = \sqrt{\mu / \rho}$.

Taking the divergence of seismic wave equation in homogeneous media, instead of the curl, yields a wave equation describing propagation of the quantity $\nabla \cdot \boldsymbol{u}$, which is the material's compression strain. The solutions of this equation, the P-waves, travel at the speed $\alpha = \sqrt{(\lambda + 2\mu) / \rho}$ that is more than twice the speed β of S-waves.

The steady-state SH waves are defined by the Helmholtz equation:

$$(\nabla^2 + k^2) \mathbf{u} = 0$$

where k is the wave number.

STONELEY WAVE

A Stoneley wave is a boundary wave (or interface wave) that typically propagates along a solid-solid interface. When found at a liquid-solid interface, this wave is also referred to as a Scholte wave. The wave is of maximum intensity at the interface and decreases exponentially away from it. It is named after the British seismologist Dr. Robert Stoneley Emeritus Professor of Seismology at the University of Cambridge, who discovered it on October 1, 1924.

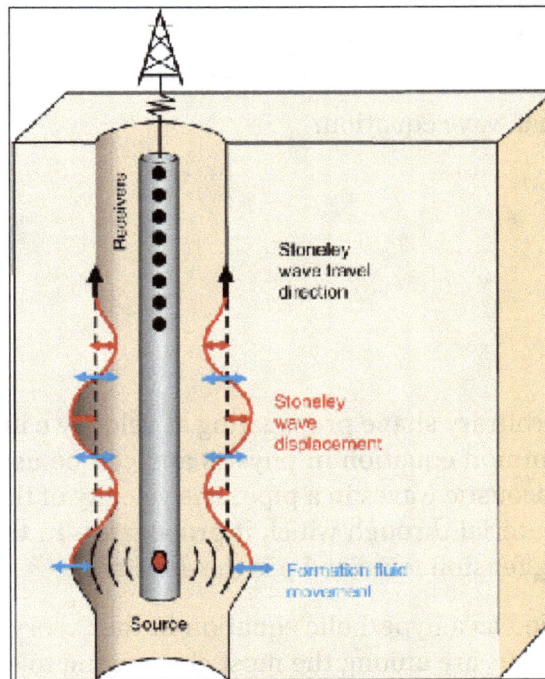

The motion of the Stoneley wave.

Occurrence and Use

Stoneley waves are most commonly generated during borehole sonic logging and vertical seismic profiling. They propagate along the walls of a fluid-filled borehole. They make up a large part of the low-frequency component of the signal from the seismic source and their attenuation is sensitive to fractures and formation permeability. Therefore, analysis of Stoneley waves can make it possible to estimate these rock properties.

Comparison to other Waves

Wave Types in Solids	Particle Vibrations
Longitudinal	Parallel to wave direction.
Transverse (Shear)	Perpendicular to wave direction.
Surface - Rayleigh	Elliptical orbit–symmetrical mode.
Plate Wave – Lamb	Component perpendicular to surface (extensional wave).
Plate Wave – Love	Parallel to plane layer, perpendicular to wave direction.
Stoneley (Leaky Rayleigh Waves)	Wave guided along interface.
Sezawa	Antisymmetric mode.

Effects of Permeability on Stoneley Wave Propagation

Permeability can influence Stoneley wave propagation in three ways. Stoneley waves can be partly reflected at sharp impedance contrasts such as fractures, lithology, or borehole diameter changes. Moreover, as formation permeability increases, Stoneley wave velocity decreases, thereby inducing dispersion. The third effect is the attenuation of Stoneley waves.

SEISMIC WAVE EQUATION

Consider the one-dimensional wave equation:

$$\frac{\partial^2 u}{\partial t^2} = c^2 \frac{\partial^2 u}{\partial x^2}$$

and its general solution:

$$u(x, t) = f(x \pm ct),$$

which represents waves of arbitrary shape propagating at velocity c in the positive and negative x directions. This is a very common equation in physics and can be used to describe, for example, the vibrations of a string or acoustic waves in a pipe. The velocity of the wave is determined by the physical properties of the material through which it propagates. In the case of a vibrating string, $c^2 = F/\rho$ where F is the string tension force and ρ is the density.

The wave equation is classified as a hyperbolic equation in the theory of linear partial differential equations. Hyperbolic equations are among the most challenging to solve because sharp features in their solutions will persist and can reflect off boundaries. Unlike, for example, the diffusion equation, solutions will be smooth only if the initial conditions are smooth. This complicates both analytical and numerical solution methods.

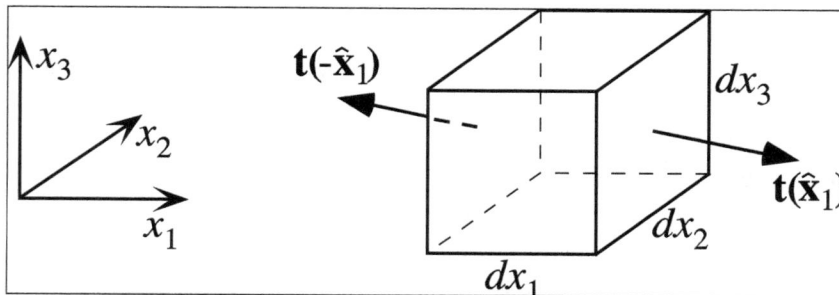

The force on the (x^2, x^3) face of an infinitesimal cube is given by $t(\hat{x}_1)$ $dx_2\, dx_3$, the product of the traction vector and the surface area.

The seismic wave equation is more complicated than equation because it is three dimensional and the link between force and displacement involves the full stress-strain relationship for an elastic solid. However, the P and S seismic wave solutions share many characteristics with the solutions to the 1-D wave equation. They involve pulses of arbitrary shape that travel at speeds determined by the elastic properties and density of the medium, and these pulses are often decomposed into harmonic wave solutions involving sine and cosine functions. Stein and Wysession provide a useful review of the 1-D wave equation as applied to a vibrating string, with analogies to seismic wave propagation in the Earth.

Momentum Equation

The stress, strain, and displacement fields were considered in static equilibrium and unchanging with time. However, because seismic waves are time-dependent phenomena that involve velocities and accelerations, we need to account for the effect of momentum. We do this by applying New-ton's law (F = ma) to a continuous medium.

Consider the forces on an infinitesimal cube in a (x_1, x_2, x_3) coordinate system. The forces on each surface of the cube are given by the product of the traction vector and the surface area. For exam-ple, the force on the plane normal to x_1 is given by:

$$
\begin{aligned}
F(\hat{x}_1) &= t(\hat{x}_1)\, dx_2\, dx_3 \\
&= \tau \hat{x}_1\, dx_2\, dx_3 \\
&= \begin{bmatrix} \tau_{11} \\ \tau_{21} \\ \tau_{31} \end{bmatrix} dx_2\, dx_3,
\end{aligned}
$$

where F is the force vector, t is the traction vector, and τ is the stress tensor. In the case of a ho-mogeneous stress field, there is no net force on the cube since the forces on opposing sides will cancel out, that is, $F(-\hat{x}1) = -F(\hat{x}1)$. Net force will only be exerted on the cube if spatial gradients are present in the stress field. In this case, the net force from the planes normal to x1 is:

$$
F(\hat{x}_1) = \frac{\partial}{\partial x_1} \begin{bmatrix} \tau_{11} \\ \tau_{21} \\ \tau_{31} \end{bmatrix} dx_1\, dx_2\, dx_3
$$

and we can use index notation and the summation convention to express the total force from the stress field on all the faces of the cube as:

$$
F_i = \sum_{j=1}^{3} \frac{\partial \tau_{ij}}{\partial x_j} dx_1\, dx_2\, dx_3
$$

$$
= \partial_j \tau_{ij}\, dx_1\, dx_2\, dx_3.
$$

The $d_j\tau_{ij}$ term is the divergence of the stress tensor (recall that summation convention means that this term is summed over j = 1, 2, 3). There may also exist a body force on the cube that acts in proportion to the volume of material, that is,

$$
F_i^{body} = f_i\, dx_1\, dx_2\, dx_3.
$$

The mass of our infinitesimal cube is given by:

$$
m = \rho\, dx_1\, dx_2\, dx_3,
$$

where ρ is the density. The acceleration of the cube is given by the second time derivative of the displacement u. Substituting F = ma and cancelling the common factor of $dx_1\, dx_2\, dx_3$, we obtain:

$$
\rho \frac{\partial^2 u_i}{\partial t^2} = \partial_j \tau_{ij} + f_i.
$$

This is the fundamental equation that underlies much of seismology. It is called the momentum equation or the equation of motion for a continuum. Each of the terms, u_i, τ_{ij} and f_i is a function of position x and time. The body force term f generally consists of a gravity term f_g and a source term f_s. Gravity is an important factor at very low frequencies in normal mode seismology, but it can generally be neglected for body- and surface-wave calculations at typically observed wavelengths. In the absence of body forces, we have the homogeneous equation of motion:

$$\rho \frac{\partial^2 u_i}{\partial t^2} = \partial_j \tau_{ij},$$

which governs seismic wave propagation outside of seismic source regions. Generating solutions for realistic Earth models is an important part of seismology; such solutions provide the predicted ground motion at specific locations at some distance from the source and are commonly termed synthetic seismograms.

If, on the other hand, we assume that the acceleration term is zero, the result is the static equilibrium equation:

$$\partial_j \tau_{ij} = -f_i.$$

in which the body forces are balanced by the divergence of the stress tensor. This equation is applicable to static deformation problems in geodesy, engineering and many other fields.

Seismic Wave Equation

In order to solve the equation we require a relationship between stress and strain so that we can express τ in terms of the displacement u. Recall the linear, isotropic stress– strain relationship,

$$\tau_{ij} = \lambda \delta_{ij} e_{kk} + 2\mu e_{ij},$$

where λ and μ are the Lamé parameters and the strain tensor is defined as:

$$e_{ij} = \frac{1}{2}\left(\partial_i u_j + \partial_j u_i\right).$$

Substituting for eij we obtain:

$$\tau_{ij} = \lambda \delta_{ij} \partial_k u_k + \mu\left(\partial_i u_j + \partial_j u_i\right).$$

Equations provide a coupled set of equations for the displacement and stress. These equations are sometimes used directly at this point to model wave propagation in computer calculations by applying finite-difference techniques. In these methods, the stresses and displacements are computed at a series of grid points in the model, and the spatial and temporal derivatives are approximated through numerical differencing. The great advantage of finite-difference schemes is their relative simplicity and ability to handle Earth models of arbitrary complexity. However, they are extremely computationally intensive and do not necessarily provide physical insight regarding the behavior of the different wave types.

In the equations that follow, we will switch back and forth between vector notation and index notation. If we substitute, we obtain:

$$\rho \frac{\partial^2 u_i}{\partial t^2} = \partial_j \left[\lambda \delta_{ij} \partial_k u_k + \mu \left(\partial_i u_j + \partial_j u_i \right) \right]$$

$$= \partial_i \lambda \partial_k u_k + \lambda \partial_i \partial_k u_k + \partial_j \mu \left(\partial_i u_j + \partial_j u_i \right) + \mu \partial_j \partial_i u_j + \mu \partial_j \partial_j u_i$$

$$= \partial_i \lambda \partial_k u_k + \partial_j \mu \left(\partial_i u_j + \partial_j u_i \right) + \lambda \partial_i \partial_k u_k + \mu \partial_i \partial_j u_j + \mu \partial_j \partial_j u_i.$$

Defining $\ddot{u} = \partial^2 u / \partial t^2$, we can write this in vector notation as:

$$\rho \ddot{u} = \nabla \lambda (\nabla \cdot u) + \nabla \mu \cdot \left[\nabla u + (\nabla u)^T \right] + (\lambda + \mu) \nabla \nabla \cdot u + \mu \nabla^2 u.$$

We now use the vector identity:

$$\nabla \times \nabla \times u = \nabla \nabla \cdot u - \nabla^2 u$$

to change this to a more convenient form. We have:

$$\nabla^2 u = \nabla \nabla \cdot u - \nabla \times \nabla \times u.$$

Substituting this, we obtain:

$$\rho \ddot{u} = \nabla \lambda (\nabla \cdot u) + \nabla \mu \cdot \left[\nabla u + (\nabla u)^T \right] + (\lambda + 2\mu) \nabla \nabla \cdot u - \mu \nabla \times \nabla \times u.$$

This is one form of the seismic wave equation. The first two terms on the right-hand side involve gradients in the Lamé parameters themselves and are nonzero whenever the material is inhomogeneous (i.e. contains velocity gradients). Most non trivial Earth models for which we might wish to compute synthetic seismograms contain such gradients. However, including these factors makes the equations very complicated and difficult to solve efficiently. Thus, most practical synthetic seismogram methods ignore these terms, using one of two different approaches.

First, if velocity is only a function of depth, then the material can be modeled as a series of homogeneous layers. Within each layer, there are no gradients in the Lamé parameters and so these terms go to zero. The different solutions within each layer are linked by calculating the reflection and transmission coefficients for waves at both sides of the interface separating the layers. The effects of a continuous velocity gradient can be simulated by considering a "staircase" model with many thin layers. As the number of layers increases, these results can be shown to converge to the continuous gradient case (more layers are needed at higher frequencies). This approach forms the basis for many techniques for computing predicted seismic motions from one-dimensional Earth models; we will term these homogeneous-layer methods. They are particularly useful for studying surface waves and low- to medium-frequency body waves. However, at high frequencies they become relatively inefficient because large numbers of layers are necessary for accurate modeling.

Second, it can be shown that the strength of these gradient terms varies as $1/\omega$, where ω is frequency, and thus at high frequencies these terms will tend to zero. This approximation is made in most ray-theoretical methods, in which it is assumed that the frequencies are sufficiently high that the $1/\omega$ terms are unimportant. However, note that at any given frequency this approximation will break down if the velocity gradients in the material become steep enough. At velocity discontinuities between regions of shallow gradients, the approximation cannot be used directly, but the solutions above and below the discontinuities can be patched together through the use of reflection and transmission coefficients. The distinction between the homogeneous-layer and ray-theoretical approaches is often important.

If we ignore the gradient terms, the momentum equation for homogeneous media becomes:

$$\rho \ddot{u} = (\lambda + 2\mu)\nabla\nabla \cdot u - \mu\nabla \times \nabla \times u.$$

This is a standard form for the seismic wave equation in homogeneous media and forms the basis for most body wave synthetic seismogram methods. However, it is important to remember that it is an approximate expression, which has neglected the gravity and velocity gradient terms and has assumed a linear, isotropic Earth model.

We can separate this equation into solutions for P-waves and S-waves by taking the divergence and curl, respectively. Taking the divergence and using the vector identity $\nabla \cdot (\nabla \times \Psi) = 0$, we obtain:

$$\frac{\partial^2 (\nabla \cdot u)}{\partial t^2} = \frac{\lambda + 2\mu}{\rho} \nabla^2 (\nabla \cdot u)$$

or,

$$\nabla^2 (\nabla \cdot u) - \frac{1}{\alpha^2} \frac{\partial^2 (\nabla \cdot u)}{\partial t^2} = 0,$$

where the P-wave velocity, α, is given by:

$$\alpha^2 = \frac{\lambda + 2\mu}{\rho}.$$

Taking the curl and using the vector identity $\nabla \times (\nabla\phi) = 0$, we obtain:

$$\frac{\partial^2 (\nabla \times u)}{\partial t^2} = -\frac{\mu}{\rho} \nabla \times \nabla \times (\nabla \times u).$$

Using the vector identity and $\nabla \cdot (\nabla \times u) = 0$, this becomes:

$$\frac{\partial^2 (\nabla \times u)}{\partial t^2} = -\frac{\mu}{\rho} \nabla^2 (\nabla \times u).$$

or,

$$\nabla^2 (\nabla \times u) - \frac{1}{\beta^2} \frac{\partial^2 (\nabla \times u)}{\partial t^2} = 0,$$

where the S-wave velocity, β, is given by:

$$\beta^2 = \frac{\mu}{\rho}.$$

We can use these equations to rewrite the elastic wave equation directly in terms of the P and S velocities:

$$\ddot{u} = \alpha^2 \, \nabla \, \nabla \cdot u \, - \, \beta^2 \, \nabla \times \nabla \times u.$$

Potentials

The displacement u is often expressed in terms of the P-wave scalar potential φ and S-wave vector potential Ψ , using the Helmholtz decomposition theorem i.e.,

$$u = \nabla \phi + \nabla \times \Psi, \, \nabla \cdot \Psi = 0..$$

We then have:

$$\nabla \cdot u = \nabla^2 \phi$$

and

$$\nabla \times u = \nabla \times \nabla \times \Psi$$
$$= \nabla \nabla \cdot \Psi - \nabla^2 \Psi$$
$$= -\nabla^2 \Psi \, \left(\text{since } \nabla \cdot \Psi = 0 \right).$$

Table: Harmonic wave parameters.

Angular frequency	ω	time⁻¹	$\omega = 2\pi f \ = \ \dfrac{2\pi}{T} = ck$
Frequency	f	time⁻¹	$f = \dfrac{\omega}{2\pi} = \dfrac{1}{T} = \dfrac{c}{\Lambda}$
Period	T	time	$T = \dfrac{1}{f} = \dfrac{2\pi}{\omega} = \dfrac{\Lambda}{c}$
Velocity	c	distance time⁻¹	$c = \dfrac{\Lambda}{T} = f\Lambda = \dfrac{\omega}{k}$
Wavelength	Λ	distance	$\Lambda = \dfrac{c}{f} = cT = \dfrac{2\pi}{k}$
Wavenumber	k	distance⁻¹	$k = \dfrac{\omega}{c} = \dfrac{2\pi}{\Lambda} = \dfrac{2\pi f}{c} = \dfrac{2\pi}{cT}$

Motivated by these equations, we require that these potentials also satisfy:

$$\nabla^2\phi \ - \ \frac{1}{\alpha^2}\frac{\partial^2\phi}{\partial t^2} = \ 0,$$

$$\nabla^2\Psi \ - \frac{1}{\beta^2}\frac{\partial^2\Psi}{\partial t^2} = \ 0.$$

After solving these equations for ϕ and Ψ, the P-wave displacement is given by the gradient of ϕ and the S-wave displacement is given by the curl of Ψ.

Plane Waves

At this point it is helpful to introduce the concept of a plane wave. This is a solution to the wave equation in which the displacement varies only in the direction of wave propagation and is constant in the directions orthogonal to the propagation direction. For example, for a plane wave traveling along the x axis, the displacement may be expressed as:

$$u(x, t) \ = \ f(t \pm x/c),$$

where c is the velocity of the wave, f is any arbitrary function (a vector function is required to express the polarization of the wave), and the waves are propagating in either the +x or −x direction. The displacement does not vary with y or z; the wave extends to infinity in these directions. If $f(t)$ is a discrete pulse, then u assumes the form of a displacement pulse traveling as a planar wavefront. More generally, displacement at position vector x for a plane wave propagating in the unit direction $\hat{s}$ may be expressed as:

$$u(x, t) \ = \ f(t - \hat{s} \cdot x/c)$$
$$= \ f(t - s \cdot x),$$

where $s = \hat{s}/c$ is the slowness vector, whose magnitude is the reciprocal of the velocity.

Since seismic energy is usually radiated from localized sources, seismic wavefronts are always curved to some extent; however, at sufficiently large distances from the source the wavefront becomes flat enough that a plane wave approximation becomes locally valid. Furthermore, many techniques for solving the seismic wave equation involve expressing the complete solution as a sum of plane waves of differing propagation angles. Often the time dependence is also removed from the equations by transforming into the frequency domain. In this case the displacement for a particular angular frequency ω may be expressed as:

$$u(x, t) \ = \ A(\omega)e^{-i\omega(t-s\cdot x)}$$
$$= \ A(\omega)e^{-i(\omega t-k\cdot x)},$$

where $k \ = \ \omega s \ = \ (\omega/c)\hat{s}$ is termed the wavenumber vector. This may be termed a monochromatic plane wave; it is also sometimes called the harmonic or steady-state plane wave solution. Other parameters used to describe such a wave are the wavenumber $k \ = \ |k| \ = \ \omega/c$, the frequency $f = \omega / (2\pi)$, the period $T = 1/f$, and the wavelength $\Lambda = cT$.

Harmonic Plane Wave Equation

What is the equation for the displacement of a 1 Hz P-wave propagating in the $+x$ direction at 6 km/s? In this case $\omega = 2\pi f$, where $f = 1\,Hz$, and thus $\omega = 2\pi$. The slowness vector is in the direction of the $x-$axis and thus $\hat{s} = \hat{x} = (1, 0, 0)$ and $s = (1/c, 0, 0) = (1/6, 0, 0)$ s/km. We can thus express as:

$$u(x, t) = u(x, t) = Ae^{-2i\pi(t-x/6)}$$

where t is in s and x is in km. As we will see in the next section, P waves are polarized in the direction of wave propagation, so $u = (ux, 0, 0)$ and we can express this more simply as:

$$u_x(x, t) = Ae^{-2i\pi(t-x/6)}$$

In general, the coefficient A is complex to permit any desired phase at $x = 0$. The real part must be taken for this equation to have a physical meaning. An alternative form is:

$$u_x(x, t) = a \cos\left[2\pi(t - x/6) - \phi\right]$$

where a is the amplitude and ϕ is the phase at $x = 0$.

Polarizations of P and S Waves

Consider plane $P-$waves propagating in the x direction. From equation we have:

$$\alpha^2 \partial_{xx}\phi = \partial_{tt}\phi.$$

Spherical Waves

A general solution to the above equation can be written as:

$$\phi = \phi_0(t \pm x/\alpha),$$

where a minus sign corresponds to propagation in the $+x$ direction and a plus sign denotes propagation in the $-x$ direction. Because $u = \nabla\phi$, we have:

$$u_x = \partial_x\phi,$$
$$u_y = 0,$$
$$u_z = 0.$$

Note that for a plane wave propagating in the x direction there is no change in the y and z directions, and so the spatial derivatives ∂_y and ∂_z are zero. For P-waves, the only displacement occurs in the direction of propagation along the x axis. Such wave motion is termed "longitudinal". Also, because $\nabla \times \nabla\phi = 0$, the motion is curl-free or "irrotational". Since $P-$waves introduce volume changes in the material $(\nabla \cdot u \neq 0)$, they can also be termed "compressional" or "dilatational". However, note that P-waves involve shearing as well as compression; this is why the P velocity is sensitive to both the bulk and shear moduli.

Now consider a plane S-wave propagating in the positive x direction. The vector potential becomes:

$$\Psi = \Psi_x(t-x/\beta)\hat{x} + \Psi_y(t-x/\beta)\hat{y} + \Psi_z(t-x/\beta)\hat{z}.$$

The displacement is:

$$u_x = (\nabla \times \Psi)_x = \partial_y\Psi_z - \partial_z\Psi_y = 0,$$
$$u_y = (\nabla \times \Psi)_y = \partial_z\Psi_x - \partial_x\Psi_z = -\partial_x\Psi_z,$$
$$u_z = (\nabla \times \Psi)_z = \partial_x\Psi_y - \partial_y\Psi_x = \partial_x\Psi_y,$$

where again we have used $\partial_y = \partial_z = 0$, thus giving:

$$u = -\partial_x\Psi_z\hat{y} + \partial_x\Psi_y\hat{z}.$$

The motion is in the y and z directions, perpendicular to the propagation direction. S – wave particle motion is often divided into two components: the motion within a vertical plane through the propagation vector (SV-waves) and the horizontal motion in the direction perpendicular to this plane (SH-waves). Because $\nabla \cdot u = \nabla \cdot (\nabla \times \Psi) = 0$, the motion is pure shear without any volume change (hence the name shear waves).

Spherical Waves

Another solution to the scalar wave equation for the P-wave potential φ is possible if we assume spherical symmetry. In spherical coordinates, the Laplacian operator is:

$$\nabla^2\phi(r) = \frac{1}{r^2}\frac{1}{\partial r}\left[r^2\frac{\partial\phi}{\partial r}\right],$$

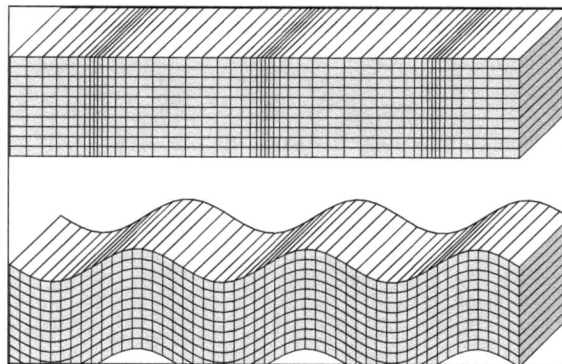

Displacements occurring from a harmonic plane P-wave (top) and S-wave (bottom) traveling horizontally across the page. S-wave propagation is pure shear with no volume change, whereas P-waves involve both a volume change and shearing (change in shape) in the material. Strains are highly exaggerated compared to actual seismic strains in the Earth.

Where we have dropped the angular derivatives because of the spherical symmetry. Using this expression, we have:

$$\frac{1}{r^2}\frac{\partial}{\partial r}\left[r^2\frac{\partial\phi}{\partial r}\right] - \frac{1}{\alpha^2}\frac{\partial^2\phi}{\partial t^2} = 0.$$

Solutions to this equation outside the point $r = 0$ may be expressed as:

$$\phi(r,\,t) = \frac{f\left(t \pm r/\alpha\right)}{r}.$$

Note that this is identical to the plane wave equation, except for the factor of $1/r$. Inward and outward propagating waves are specified by the $+$ and $-$ signs respectively. Since this expression is usually used to model waves radiating away from a point source, the inward propagating solution is normally ignored. In this case the $1/r$ term represents a decay in the wave amplitude with range.

Equation is not a valid solution at $r = 0$. However, it can be shown that the solution to the inhomogeneous wave equation:

$$\nabla^2\phi(r) - \frac{1}{\alpha^2}\frac{\partial^2\varphi}{\partial t^2} = -4\pi\delta(r)f(t),$$

where the delta function $\delta(r)$ is zero everywhere except $r = 0$ and has a volume integral of one. The factor $4\pi\delta(r)f(t)$ represents the source–time function at the origin.

Methods for Computing Synthetic Seismograms

A large part of seismology involves devising and implementing techniques for computing synthetic seismograms for realistic Earth models. In general, our goal is to calculate what would be recorded by a seismograph at a specified receiver location, given an exact specification of the seismic source and the Earth model through which the seismic waves propagate. This is a well-defined forward problem that, in principle, can be solved exactly. However, errors in the synthetic seismograms often occur in practical applications. These inaccuracies can be separated into two parts:

- Inaccuracies arising from approximations in the theory used to compute the synthetic seismograms. Examples of this would include many applications of ray theory which do not properly account for head waves, diffracted waves, or the coupling between different waves types at long periods. Another computational error is the grid dispersion that occurs in most finite difference schemes.

- Errors caused by using a simplified Earth or source model. In this case the synthetic seismogram may be exact for the simplied model, but the model is an inadequate representation of the real problem. These simplifications might be necessary in order to apply a particular numerical technique, or might result from ignorance of many of the details of the model. Examples would include the use of 1-D models that do not fully account for 3-D structure, the assumption of a point source rather than a finite rupture, and neglecting the effects of attenuation or anisotropy in the calculations.

The first category of errors may be addressed by applying a more exact algorithm, although in practice limits on computing resources may prevent achieving the desired accuracy in the case of complicated models. The second category is more serious because often one simply does not know the properties of the Earth well enough to be able to model every wiggle in the observed seismograms. This is particularly true at high frequencies (0.5 Hz and above). For teleseismic arrivals, long-period body waves (15–50 s period) and surface waves (40–300 s period) can usually be fit well with current Earth models, whereas the coda of high-frequency body wave arrivals can only be modeled statistically (fitting the envelope function but not the individual wiggles).

Because of the linearity of the problem and the superposition principle (in which distributed sources can be described as the sum of multiple point sources), there is no great difficulty in modeling even very complicated sources (inverting for these sources, is, of course, far more difficult, but here we are only concerned with the forward problem). If the source can be exactly specified, then computing synthetics for a distributed source is only slightly more complicated than for a simple point source. By far the most difficult part in computing synthetic seismgrams is solving for the propogation effects through realistic velocity structures. Only for a few grossly simplified models (e.g. whole space or half-spaces) are analytical solutions possible.

The part of the solution that connects the force distribution at the source with the displacements at the receiver is termed the elastodynamic Green's function. Computation of the Green's function is the key part of the synthetic seismogram calculation because this function must take into account all of the elastic properties of the material and the appropriate boundary conditions.

There are a large number of different methods for computing synthetic seismograms. Most of these fall into the following categories:

- Finite-difference and finite-element methods that use computer power to solve the wave equation over a discrete set of grid points or model elements. These have the great advantage of being able to handle models of arbitrary complexity. Their computational cost grows with the number of required grid points; more points are required for 3-D models (vs. 2-D) and for higher frequencies.

- Ray-theoretical methods in which ray geometries are explicitly specified and ray paths are computed. These methods include simple (or geometrical) ray theory, WKBJ, and so-called "generalized" ray theory. They are most useful at high frequencies for which the ray-theoretical approximation is most accurate. They are most simply applied to 1-D Earth models but can be generalized to 3-D models.

- Homogeneous layer methods in which the model consists of a series of horizontal layers with constant properties within each layer. Matrix methods are then used to connect the solutions between layers. Examples of this approach include "reflectivity" and "wavenumber integration". These methods yield an exact solution but can become moderately computationally intensive at high frequencies because a large number of layers are required to accurately simulate continuous velocity gradients. Unlike finite-difference and ray-theoretical methods, homogeneous-layer techniques are restricted to 1-D Earth models. However, spherically symmetric models can be computed using the Flat Earth Transformation.

- Normal mode summation methods in which the standing waves (eigenvectors) of the spherical Earth are computed and then summed to generate synthetic seismograms. This is the most natural and complete way to compute synthetic seismograms for the spherical Earth, but is computationally intensive at high frequencies. Generalization to 3-D Earth models requires including coupling between modes; this is generally done using asymptotic approximations and greatly increases the complexity of the algorithm.

There is no single "best" way to compute synthetic seismograms as each method has its own advantages and disadvantages. The method of choice will depend upon the particular problem to be addressed and the available computer power; thus it is useful to be aware of the full repertoire of techniques.

ADAMS–WILLIAMSON EQUATION

The Adams–Williamson equation, named after L. H. Adams and E. D. Williamson, is an equation used to determine density as a function of radius, more commonly used to determine the relation between the velocities of seismic waves and the density of the Earth's interior. Given the average density of rocks at the Earth's surface and profiles of the P-wave and S-wave speeds as function of depth, it can predict how density increases with depth. It assumes that the compression is adiabatic and that the Earth is spherically symmetric, homogeneous, and in hydrostatic equilibrium. It can also be applied to spherical shells with that property. It is an important part of models of the Earth's interior such as the Preliminary reference Earth model (PREM).

Theory

The two types of seismic body waves are compressional waves (P-waves) and shear waves (S-waves). Both have speeds that are determined by the elastic properties of the medium they travel through, in particular the bulk modulus K, the shear modulus μ, and the density ρ. In terms of these parameters, the P-wave speed v_p and the S-wave speed v_s are:

$$v_p = \sqrt{\frac{K + (4/3)\mu}{\rho}}$$

$$v_s = \sqrt{\frac{\mu}{\rho}}.$$

These two speeds can be combined in a seismic parameter:

$$\Phi = v_p^2 - \frac{4}{3}v_s^2 = \frac{K}{\rho}.$$

The definition of the bulk modulus,

$$K = -V\frac{dP}{dV},$$

is equivalent to:

$$K = \rho\frac{dP}{d\rho}.$$

Suppose a region at a distance r from the Earth's center can be considered a fluid in hydrostatic equilibrium, it is acted on by gravitational attraction from the part of the Earth that is below it and pressure from the part above it. Also suppose that the compression is adiabatic (so thermal expansion does not contribute to density variations). The pressure $P(r)$ varies with r as:

$$\frac{dP}{dr} = -\rho(r)g(r),$$

where $g(r)$ is the gravitational acceleration at radius r.

If equations are combined, we get the Adams–Williamson equation:

$$\frac{d\rho}{dr} = -\frac{\rho(r)g(r)}{\Phi(r)}.$$

This equation can be integrated to obtain:

$$\ln\left(\frac{\rho}{\rho_0}\right) = -\int_{r_0}^{r} \frac{g(r)}{\Phi(r)} dr,$$

where r_0 is the radius at the Earth's surface and ρ_0 is the density at the surface. Given ρ_0 and profiles of the P- and S-wave speeds, the radial dependence of the density can be determined by numerical integration.

SEISMIC WAVE PROPAGATION

The full elastic seismic wavefield that propagates through an isotropic Earth consists of a P-wave component and two shear (SV and SH) wave components. Marine air guns and vertical onshore sources produce reflected wavefields that are dominated by P and SV modes. Much of the SV energy in these wavefields is created by P-to-SV-mode conversions when the downgoing P wavefield arrives at stratal interfaces at nonnormal angles of incidence. Horizontal-dipole sources can create strong SH modes in onshore programs. No effective seismic horizontal-dipole sources exist for marine applications.

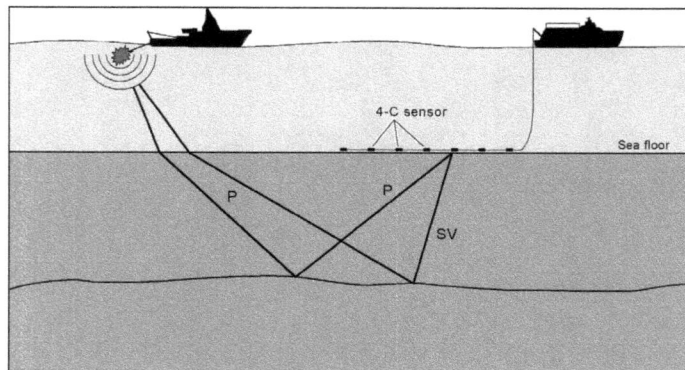

Basic elements of ocean-bottom cable data acquisition. P denotes a seismic
compressional wave; SV is a converted shear mode.

Seismic Wave Model

A principal difference among P, SV, and SH wavefields is the manner in which they cause rock particles to oscillate. Figure illustrates the relationships between propagation direction and particle-displacement direction for these three wave modes. A compressional wave causes rock particles to oscillate in the direction that the wavefront is propagating. In other words, a P-wave particle displacement vector is perpendicular to its associated P-wave wavefront. In contrast, SV and SH waves cause rock particles to oscillate perpendicular to the direction that the wavefront is moving, with the SH and SV displacement vectors orthogonal to each other. A shear-wave particle-displacement vector is thus tangent to its associated wavefront. In a flat-layered isotropic Earth, the SH displacement vector is parallel to stratal bedding, and SV displacement is in the plane that is perpendicular to bedding.

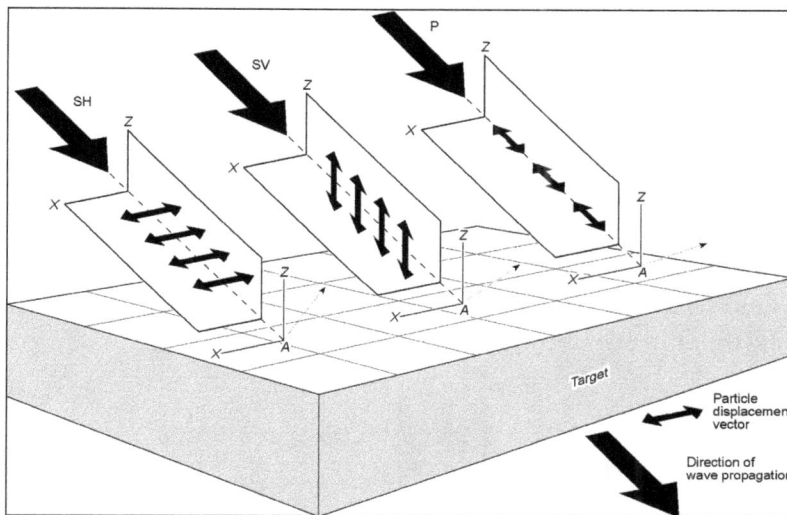

Distinction between the three components of an elastic wavefield.

To create optimal images of subsurface targets, a seismic wavefield must be segregated into its P, SV, and SH component parts so that a P-wave image can be made that has minimal contamination from interfering SV and SH modes. Likewise, an SV image must have no interfering P and SH modes, and an SH image must be devoid of P and SV contamination.

A P-wave travels at velocity V_p in consolidated rocks, which is approximately two times faster than velocity V_s of either the SH or SV wave. In carbonates, the velocity ratio (V_p/V_s) tends to be approximately 1.7 or 1.8. In siliciclastics, V_p/V_s varies from approximately 1.6 in hard sandstones to approximately 3 in some shales. This velocity difference aids in separating interfering P and S wave modes during data processing. An equally powerful technique for separating a seismic wavefield into its component parts is to use data-processing techniques that concentrate on the distinctions in the particle displacements associated with the P, SH, and SV modes.

The P, SH, and SV particle displacements form an orthogonal coordinate system. The fundamental requirement of multicomponent seismic imaging is that reflection wavefields must be recorded with orthogonal 3C sensors that allow these P, SH, and SV particle motions to be recognized. To date, most exploration seismic data have been recorded with single-component sensors that emphasize P-wave modes and do not capture SH or SV wave modes.

Body Waves and Surface Waves

Seismic wavefields propagate through the Earth in two ways: Body waves and surface waves. Body waves propagate in the interior (body) of the Earth and illuminate deep geologic targets. These waves generate the reflected P, SH, and SV signals that are needed to evaluate prospects and to characterize reservoirs. Reflected (or scattered) body waves are the fundamental signals sought in seismic data-acquisition programs.

Surface waves travel along the Earth/air interface and do not illuminate geologic targets in the interior of the Earth. Surface waves are noise modes that overlay the desired body-wave reflections. Surface waves can be a serious problem in onshore seismic surveys. Surface waves do not affect towed-cable marine data because they require some shear-wave component to propagate, and shear waves cannot propagate along the air/water interface. An exception in the marine case is sometimes encountered when data are recorded with ocean-bottom sensors (OBS) because interface waves can propagate along the water/sediment boundary and become a type of surface-wave noise that degrades OBS marine seismic data.

There are two principal surface waves: Love waves and Rayleigh waves. Love waves are an SH-mode surface wave and do not affect conventional P-wave seismic data. Love waves are a serious noise mode only when the objective is to record reflected SH wavefields. The more common surface wave is the Rayleigh wave, which combines P and SV motions and is referred to as ground roll on P-wave seismic field records. Love waves create particle displacements in the horizontal plane; Rayleigh wave displacements are in the vertical plane.

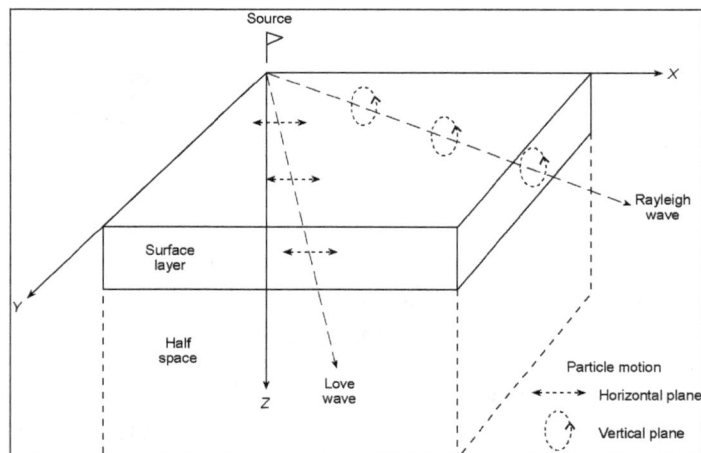

Particle motions produced by the two principal seismic surface-wave
noise modes: the Love wave and the Rayleigh wave.

Much of the field effort in onshore seismic programs concentrates on designing and deploying receiver arrays that can attenuate horizontally traveling surface waves (ground-roll noise) and, at the same time, amplify upward-traveling reflection signals. The most effective field technique is to deploy 10, 12, 16, or more geophones at a uniform spacing at each receiver station so that the distance from the first geophone to the last geophone is the same as the dominant wavelength of the ground-roll event. All geophone responses are then summed to create a single output response at that receiver station. The idea is to create a sensor array length such that half of the geophones are moving up and half are moving down as the horizontally traveling ground roll passes the receiver station. The summed output of the geophone array is essentially zero because of the passage of the ground-roll event. In contrast, upward-traveling reflections arriving at this same receiver array are not attenuated because such events cause all geophones to move up and down in unison. The summed output of the array for an upward-traveling reflection wavefield is thus a strong voltage signal.

Seismic Impedance

The concept of acoustic (or seismic) impedance is critical to understanding seismic reflectivity. Seismic impedance controls the seismic reflection process in the sense that seismic energy is reflected only at rock interfaces in which there are changes in impedance across the interface. Seismic impedance is defined as:

$$I = pV,$$

where I = impedance, ρ = the bulk density of the rock, and V = the velocity of seismic wave propagation through the rock. V is set to V_p if the wave mode of interest is a P-wave; it is set to V_s if S-wave reflectivity is being considered. Any alteration in rock properties that causes ρ and/or V to change can be the genesis of a seismic reflection event; therefore, areal and vertical variations in seismic reflectivity can be used to infer spatial distributions of rock types and porosity trends.

Reflection Coefficients

Seismic reflectivity is best explained with a simple two-layer Earth model in which Layer 1 is above Layer 2. The seismic reflection coefficient, R, for a downgoing particle-velocity wave mode that arrives perpendicular to the interface between the two layers is:

$$R = \frac{I_1 - I_2}{I_1 + I_2} = \frac{\rho_1 V_1 - \rho_2 V_2}{\rho_1 V_1 + \rho_2 V_2},$$

A negative algebraic sign has to be inserted on the right side of equation if the downgoing wavefield is a pressure wavefield (hydrophone measurement) rather than a particle-velocity wavefield (geophone measurement). The velocity parameters, V_1 and V_2, are P-wave velocities if P-wave reflectivity is being calculated; they are S-wave velocities if S-wave reflectivity is to be determined. At any interface, R can be positive, negative, or zero, depending on the impedance contrast $(\rho_1 V_1 - \rho_2 V_2)$ across the interface.

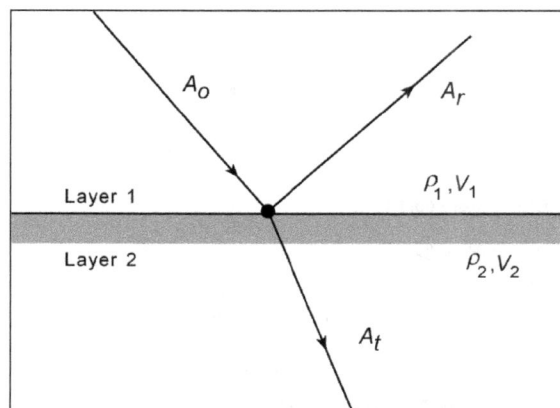

Two-layer Earth model describing seismic reflectivity parameters.

A seismic reflection/transmission process is indicated in Figure by the raypaths labeled A_o, A_r, and A_t. For nonnormal incidence angles, the expression for reflection coefficient involves trigonometric functions that ensure that horizontal slowness (the inverse of horizontal velocity) is conserved and is a more complex expression. The seismic reflection, A_r, is given by:

$$A_r = RA_o,$$

and the transmitted seismic event, A_t, is given by:

$$A_t = (1 + R)A_o$$

The magnitude and algebraic sign of A_r depends on R and, in turn, the basic control on R is the variation of impedance ρV across the interface.

Two types of petrophysical properties control the value of acoustic impedance in individual rock layers: elastic properties of the rock matrix and properties of the fluid in the pore spaces of the rock. P-waves travel through elastic materials and fluids; thus, any change in either the rock matrix (such as a change in mineralogy or porosity) or in the type of fluid occupying the pore spaces will create a discontinuity in the P-wave seismic impedance of the rock system.

Figure illustrates the relationships between petrophysical conditions that occur at an impendance boundary and the existence of P and S reflections at that boundary. A P-wave reflection will occur at boundaries at which there is a change in either the rock matrix or the pore fluid, or both. In constrast, S-waves are not affected by changes in pore fluid or are only weakly affected. Consequently, a change in the properties of the rock matrix can create a reflecting boundary for S-waves, but a change in pore fluid will create only a small (usually negligible) S-wave reflection boundary. If a small, nonzero S-wave reflection coefficient occurs at a fluid boundary, that reflection coefficient usually exists because the bulk density of the rock system varies across the fluid boundary.

ROCK AND FLUID PROPERTIES THAT CREATE P AND S REFLECTIONS		
Boundary conditions	P-reflection	S-reflection
Rock matrix 1, Pore fluid 1 //////////// Rock matrix 1, Pore fluid 2	Yes	No (may be weak)
Rock matrix 1, Pore fluid 1 //////// Rock matrix 2, Pore fluid 1	Yes	Yes
Rock matrix 1, Pore fluid 1 /////// Rock matrix 2, Pore fluid 2	Yes	Yes

Relationships between petrophysical conditions that occur at an impedance boundary and the existence of P and S reflections at that boundary.

These wave physics provide valuable geologic insights when both P and S reflection data are acquired across a prospect area. When P and S reflections occur at the same depth coordinate, the reflecting boundary at that depth is associated with a change in the rock matrix (that is, with a lithological change). There may or may not be a change in pore fluid at that boundary. When a P reflection occurs at a boundary but there is no S reflection, that boundary quite likely marks a change in pore fluid and not a change in rock matrix. That is, the lithology probably does not change at that depth, but the type of pore fluid does.

References

- Seismic-wave, science: britannica.com, Retrieved 05 July, 2019

- Surface-waves, seismic-science-and-site-influences-earthquake-energy: seismicresilience.org.nz, Retrieved 10 August, 2019

- Goldstein, R.V.; Gorodtsov, V.A.; Lisovenko, D.S. (2014). "Rayleigh and Love surface waves in isotropic media with negative Poisson's ratio". Mechanics of Solids. 49 (4): 422–434. doi:10.3103/S0025654414040074

- Sheikhhassani, Ramtin (2013). "Scattering of a plane harmonic SH wave by multiple layered inclusions". Wave Motion. 51 (3): 517–532. doi:10.1016/j.wavemoti.2013.12.002

- Surface-wave-methods, archive-geophysics: archive.epa.gov, Retrieved 08 April, 2019

- Seismic-wave-propagation: petrowiki.org, Retrieved 26 January, 2019

Seismic Methods and Measurements

Seismic methods include reflection seismology, seismic refraction, isoseismal map, passive seismic, plus minus method, etc. Seismometer, seismogram, geophone, accelerograph, etc. are some of the seismic instruments. This chapter has been carefully written to provide an easy understanding of these seismic methods and instruments.

Seismic methods are the most commonly conducted geophysical surveys for engineering investigations. Seismic refraction provides engineers and geologists with the most basic of geologic data via simple procedures with common equipment.

Any mechanical vibration is initiated by a source and travels to the location where the vibration is noted. These vibrations are seismic waves. The vibration is merely a change in the stress state due to a disturbance. The vibration emanates in all directions that support displacement. The vibration readily passes from one medium to another and from solids to liquids or gasses and in reverse. A vacuum cannot support mechanical vibratory waves, while electromagnetic waves can be transmitted through a vacuum. The direction of travel is called the ray, ray vector, or ray path. Since a source produces motion in all directions the locus of first disturbances will form a spherical shell or wave front in a uniform material.

Wave Theory

A seismic disturbance moves away from a source location; the locus of points defining the expanding disturbance is termed the wavefront. Any point on a wavefront acts as a new source and causes displacements in surrounding positions. The vector normal to the wavefront is the ray path through that point, and is the direction of propagation. Upon striking a boundary between differing material properties, wave energy is transmitted, reflected, and converted. The properties of the two media and the angle at which the incident ray path strikes will determine the amount of energy reflected off the surface, refracted into the adjoining material, lost as heat, and changed to other wave types.

An S-wave in rock approaching a boundary of a lake will have an S-wave reflection, a P-wave reflection, and a likely P-wave refraction into the lake water (depending on the properties and incident angle). Since the rock-water boundary will displace, energy will pass into the lake, but the water cannot support an S-wave. The reflected S-wave departs from the boundary at the same angle normal to the boundary as the arriving S-wave struck. In the case of a P-wave incident on a boundary between two rock types (of differing elastic properties), there may be little conversion to S-waves. Snell's Law provides the angles of reflection and refraction for both the P- and S-waves. In the rock on the source side (No. 1), the velocities are V_{P_1} and V_{S_1}; the second rock material (No. 2) has properties of V_{P2} and V_{S2}. Then for the incident Pwave (P_ii), Snell's Law provides the angles of reflections in rock No. 1 and refraction in rock No. 2 as:

$$\frac{\sin \alpha_{P1i}}{V_{P1i}} = \frac{\sin \alpha_{P1}}{V_{P1}} = \frac{\sin \alpha_{S1}}{V_{S1}} = \frac{\sin \alpha_{P2}}{V_{P2}} = \frac{\sin \alpha_{S2}}{V_{S2}}$$

The second and third terms of equation are reflections within material No. 1; the fourth and fifth terms are refractions into medium No. 2. Note that none of the angles can exceed 90 degrees, since none of the sin terms can be over 1.0, and $\alpha_{P1i} = \alpha_{P1}$.

Two important considerations develop from understanding equation. First is the concept of critical refraction. If rock No. 1 has a lower velocity than rock No. 2 or $V_{P1} < V_{P2}$, then from equation, $\sin \alpha_{P2} > \sin \alpha_{P1i}$ and the refracted $\alpha_{P2} > \alpha_{P1i}$, the incident angle. Yet $\sin \alpha_{P2}$ cannot exceed 1.0. The critical incident angle causes the refraction to occur right along the boundary at $90°$ from the normal to the surface. The critical angle is that particular incident angle such that sine $\alpha_{P2} = 1.0$ and $\alpha_{P2} = 90°$, or $\alpha_{(P1i)cr} = \sin\text{-}1(V_{P1}/V_{P2})$. Secondly, any incident angle $> \alpha_{(P1i)cr}$ from the normal will cause total reflection back into the source-side material, since $\sin \alpha_{P2}$ Ý 1.0. For the latter case, all the P-wave energy will be retained in medium No. 1.

Other wave phenomena occur in the subsurface. Diffractions develop at the end of sharp boundaries. Scattering occurs due to inhomogeneities within the medium. As individual objects shrink in size, their effect on scatter is reduced. Objects with mean dimensions smaller than one- fourth of the wavelength will have little effect on the wave. Losses of energy or attenuation occur with distance of wave passage. Higher frequency waves lose energy more rapidly than waves of lower frequencies, in general.

The wave travels outward from the source in all directions supporting displacements. Energy dissipation is a function of the distance travelled, as the wave propagates away from the source. At boundaries, the disturbance passes into other media. If a wave can pass from a particular point A to another point B, Fermat's principle states that the ray path taken is the one requiring the minimum amount of time. In crossing boundaries of media with different properties, the path will not be the shortest distance (a straight line) due to refraction. The actual ray path will have the shortest travel time. Since every point on a wavefront is a new source, azimuths other than that of the fastest arrival will follow paths to other locations for the ever-expanding wave.

Data Acquisition

Digital electronics have continued to allow the production of better seismic equipment. Newer equipment is hardier, more productive, and able to store greater amounts of data. The choice of seismograph, sensors (geophones), storage medium, and source of the seismic wave depend on the survey being undertaken. The sophistication of the survey, in part, governs the choice of the equipment and the field crew size necessary to obtain the measurements. Costs rise as more elaborate equipment is used. However, there are efficiencies to be gained in proper choice of source, number of geophone emplacements for each line, crew size, channel capacity of the seismograph, and requirements of the field in terrain type and cultural noise.

The seismic source may be a hammer striking the ground or an aluminum plate or weighted plank, drop weights of varying sizes, rifle shot, a harmonic oscillator, waterborne mechanisms, or explosives. The energy disturbance for seismic work is most often called the "shot," an archaic term from petroleum seismic exploration. Reference to the "shot" does not necessarily mean an explosive or rifle source was used. The type of survey dictates some source parameters. Smaller mass, higher frequency sources are preferable. Higher frequencies give shorter wavelengths and more precision in choosing arrivals and estimating depths. Yet, sufficient energy needs to be transmitted to obtain a strong return at the end of the survey line. The type of source for a particular survey is usually known prior to going into the field. A geophysical contractor normally should be given latitude in selecting or changing the source necessary for the task. The client should not hesitate in placing limits on the contractor's indiscriminate use of some sources. In residential or industrial areas, perhaps the maximum explosive ge should be limited. The depth of drilling shot holes for explosives or rifle shots may need to be limited; contractors should be cautious not to exceed requirements of permits, utility easements, and contract agreements.

The sensor receiving seismic energy is the geophone (hydrophone in waterborne surveys) or phone. These sensors are either accelerometers or velocity transducers, and convert ground movement into a voltage. Typically, the amplification of the ground is many orders of magnitude, but accomplished on a relative basis. The absolute value of particle acceleration cannot be determined, unless the geophones are calibrated.

Most geophones have vertical, single-axis response to receive the incoming waveform from beneath the surface. Some geophones have horizontal-axis response for S-wave or surface wave assessments. Triaxial phones, capable of measuring absolute response, are used in specialized surveys. Geophones are chosen for their frequency band response.

The line, spread, or string of phones may contain one to scores of sensors depending on the type of survey. The individual channel of recording normally will have a single phone. Multiple phones per channel may aid in reducing wind noise or air blast or in amplifying deep reflections.

The equipment that records input geophone voltages in a timed sequence is the seismograph. Current practice uses seismographs that store the channels' signals as digital data at discrete time. Earlier seismographs would record directly to paper or photographic film. Stacking, inputting, and processing the vast volumes of data and archiving the information for the client virtually require digital seismographs. The seismograph system may be an elaborate amalgam of equipment to trigger or sense the source, digitize geophone signals, store multichannel data, and provide some level of processing display. Sophisticated seismograph equipment is not normally required for engineering and environmental surveys. One major exception is the equipment for sub-bottom surveys or nondestructive testing of pavements.

Data processing of seismic information can be as simple as tabular equations for seismic refraction. Processing is normally the most substantial matter the geophysicists will resolve, except for the interpretation.

A portion of the seismic energy striking an interface between two differing materials will be reflected from the interface. The ratio of the reflected energy to incident energy is called the reflection coefficient. The reflection coefficient is defined in terms of the densities and seismic velocities of the two materials as:

$$R = \frac{(p_{b2}V_2 - p_{b1}V_1)}{(p_{b2}V_2 + p_{b1}V_1)},$$

where,

R = reflection coefficient.

p_{b1}, p_{b2} = densities of the first and second layers, respectively.

V_1, V_2 = seismic velocities of the first and second layers, respectively.

Modern reflection methods can ordinarily detect isolated interfaces whose reflection coefficients are as small as 0.02.

REFLECTION SEISMOLOGY

Reflection seismology (or seismic reflection) is a method of exploration geophysics that uses the principles of seismology to estimate the properties of the Earth's subsurface from reflected seismic waves. The method requires a controlled seismic source of energy, such as dynamite or Tovex blast, a specialized air gun or a seismic vibrator, commonly known by the trademark name Vibroseis. Reflection seismology is similar to sonar and echolocation.

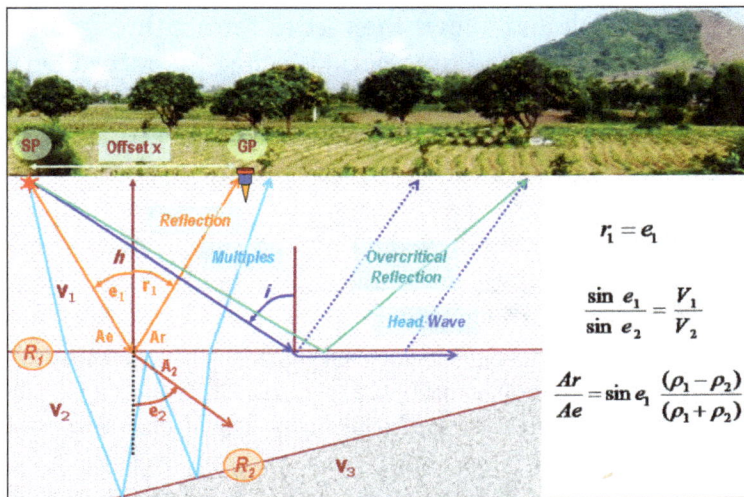

Seismic Reflection Outlines.

Seismic waves are mechanical perturbations that travel in the Earth at a speed governed by the acoustic impedance of the medium in which they are travelling. The acoustic (or seismic) impedance, Z, is defined by the equation:

$$Z = V\rho,$$

where V is the seismic wave velocity and ρ is the density of the rock.

When a seismic wave travelling through the Earth encounters an interface between two materials with different acoustic impedances, some of the wave energy will reflect off the interface and some will refract through the interface. At its most basic, the seismic reflection technique consists of generating seismic waves and measuring the time taken for the waves to travel from the source, reflect off an interface and be detected by an array of receivers (or geophones) at the surface. Knowing the travel times from the source to various receivers, and the velocity of the seismic waves, a geophysicist then attempts to reconstruct the pathways of the waves in order to build up an image of the subsurface.

In common with other geophysical methods, reflection seismology may be seen as a type of inverse problem. That is, given a set of data collected by experimentation and the physical laws that apply to the experiment, the experimenter wishes to develop an abstract model of the physical system being studied. In the case of reflection seismology, the experimental data are recorded seismograms, and the desired result is a model of the structure and physical properties of the Earth's crust. In common with other types of inverse problems, the results obtained from reflection seismology are usually not unique (more than one model adequately fits the data) and may be sensitive to relatively small errors in data collection, processing, or analysis. For these reasons, great care must be taken when interpreting the results of a reflection seismic survey.

Reflection Experiment

The general principle of seismic reflection is to send elastic waves (using an energy source such as dynamite explosion or Vibroseis) into the Earth, where each layer within the Earth reflects a portion of the wave's energy back and allows the rest to refract through. These reflected energy waves are recorded over a predetermined time period (called the record length) by receivers that detect the motion of the ground in which they are placed. On land, the typical receiver used is a small, portable instrument known as a geophone, which converts ground motion into an analogue electrical signal. In water, hydrophones are used, which convert pressure changes into electrical signals. Each receiver's response to a single shot is known as a "trace" and is recorded onto a data storage device, then the shot location is moved along and the process is repeated. Typically, the recorded signals are subjected to significant amounts of signal processing before they are ready to be interpreted and this is an area of significant active research within industry and academia. In general, the more complex the geology of the area under study, the more sophisticated are the techniques required to remove noise and increase resolution. Modern seismic reflection surveys contain large amount of data and so require large amounts of computer processing, often performed on supercomputers or computer clusters.

Reflection and Transmission at Normal Incidence

When a seismic wave encounters a boundary between two materials with different acoustic impedances, some of the energy in the wave will be reflected at the boundary, while some of the energy will be transmitted through the boundary. The amplitude of the reflected wave is predicted by multiplying the amplitude of the incident wave by the seismic reflection coefficient R, determined by the impedance contrast between the two materials.

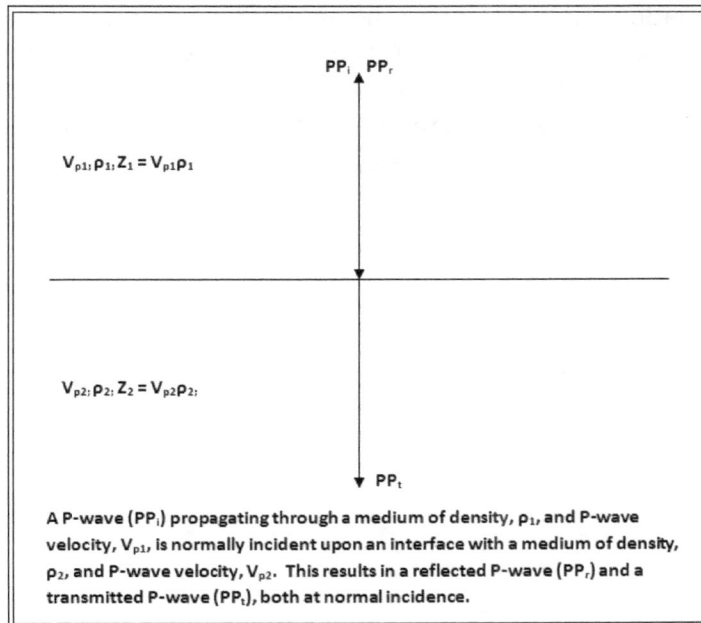

A P-wave (PP$_i$) propagating through a medium of density, ρ_1, and P-wave velocity, V$_{p1}$, is normally incident upon an interface with a medium of density, ρ_2, and P-wave velocity, V$_{p2}$. This results in a reflected P-wave (PP$_r$) and a transmitted P-wave (PP$_t$), both at normal incidence.

P-wave reflects off an interface at normal incidence.

For a wave that hits a boundary at normal incidence (head-on), the expression for the reflection coefficient is simply:

$$R = \frac{Z_2 - Z_1}{Z_2 + Z_1},$$

where Z_1 and Z_2 are the impedance of the first and second medium, respectively.

Similarly, the amplitude of the incident wave is multiplied by the transmission coefficient to predict the amplitude of the wave transmitted through the boundary. The formula for the normal-incidence transmission coefficient is:

$$T = 1 - R = \frac{2Z_1}{(Z_2 + Z_1)}.$$

As the sum of the squares of amplitudes of the reflected and transmitted wave has to be equal to the square of amplitude of the incident wave, it is easy to show that:

$$Z_1(1 - R^2) = \frac{Z_1(Z_2 + Z_1)^2 - Z_1(Z_2 - Z_1)^2}{(Z_2 + Z_1)^2} = \frac{4Z_2 Z_1^2}{(Z_2 + Z_1)^2} = Z_2 T^2.$$

By observing changes in the strength of reflectors, seismologists can infer changes in the seismic impedances. In turn, they use this information to infer changes in the properties of the rocks at the interface, such as density and elastic modulus.

Reflection and Transmission at Non-normal Incidence

The situation becomes much more complicated in the case of non-normal incidence, due to mode conversion between P-waves and S-waves, and is described by the Zoeppritz equations. In 1919, Karl

Zoeppritz derived 4 equations that determine the amplitudes of reflected and refracted waves at a planar interface for an incident P-wave as a function of the angle of incidence and six independent elastic parameters. These equations have 4 unknowns and can be solved but they do not give an intuitive understanding for how the reflection amplitudes vary with the rock properties involved.

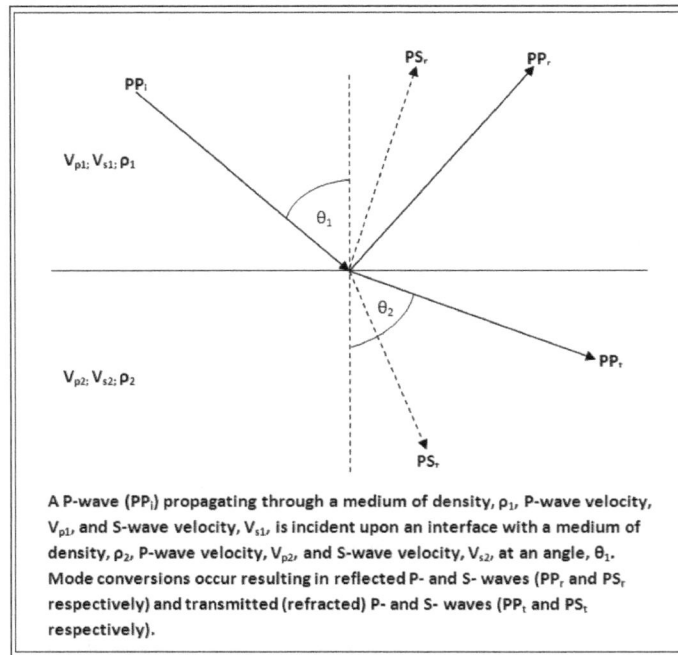

A P-wave (PP$_i$) propagating through a medium of density, ρ_1, P-wave velocity, V$_{p1}$, and S-wave velocity, V$_{s1}$, is incident upon an interface with a medium of density, ρ_2, P-wave velocity, V$_{p2}$, and S-wave velocity, V$_{s2}$, at an angle, θ_1. Mode conversions occur resulting in reflected P- and S- waves (PP$_r$ and PS$_r$ respectively) and transmitted (refracted) P- and S- waves (PP$_t$ and PS$_t$ respectively).

Diagram showing the mode conversions that occur when a P-wave reflects off an interface at non-normal incidence.

The reflection and transmission coefficients, which govern the amplitude of each reflection, vary with angle of incidence and can be used to obtain information about (among many other things) the fluid content of the rock. Practical use of non-normal incidence phenomena, known as AVO has been facilitated by theoretical work to derive workable approximations to the Zoeppritz equations and by advances in computer processing capacity. AVO studies attempt with some success to predict the fluid content (oil, gas, or water) of potential reservoirs, to lower the risk of drilling unproductive wells and to identify new petroleum reservoirs. The 3-term simplification of the Zoeppritz equations that is most commonly used was developed in 1985 and is known as the "Shuey equation". A further 2-term simplification is known as the "Shuey approximation", is valid for angles of incidence less than 30 degrees (usually the case in seismic surveys) and is given below:

$$R(\theta) = R(0) + G\sin^2\theta$$

where $R(0)$ = reflection coefficient at zero-offset (normal incidence); G = AVO gradient, describing reflection behaviour at intermediate offsets and θ = angle of incidence. This equation reduces to that of normal incidence at $\theta = 0$.

Interpretation of Reflections

The time it takes for a reflection from a particular boundary to arrive at the geophone is called the travel time. If the seismic wave velocity in the rock is known, then the travel time may be used to estimate the depth to the reflector. For a simple vertically traveling wave, the travel time t from the surface to the reflector and back is called the Two-way Time (TWT) and is given by the formula:

$$t = 2\frac{d}{V},$$

where d is the depth of the reflector and V is the wave velocity in the rock.

A series of apparently related reflections on several seismograms is often referred to as a reflection event. By correlating reflection events, a seismologist can create an estimated cross-section of the geologic structure that generated the reflections. Interpretation of large surveys is usually performed with programs using high-end three-dimensional computer graphics.

Sources of Noise

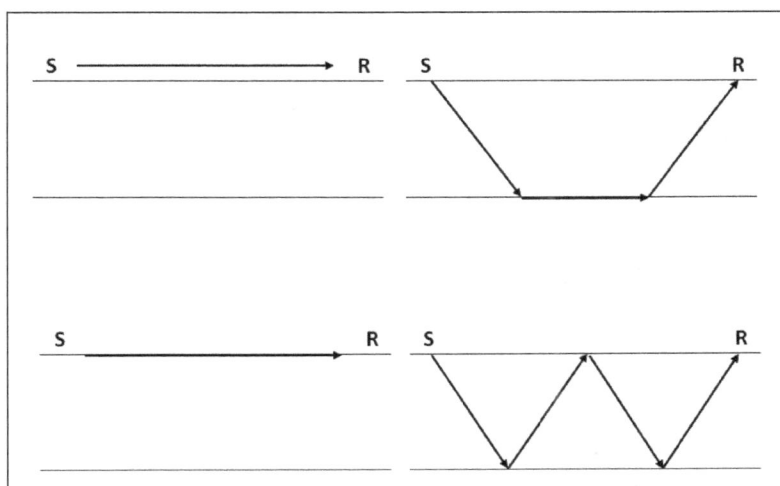

Sources of noise on a seismic record. Top-left: air wave; top-right: head wave; bottom-left: surface wave; bottom-right: multiple.

In addition to reflections off interfaces within the subsurface, there are a number of other seismic responses detected by receivers and are either unwanted or unneeded:

Air Wave

The airwave travels directly from the source to the receiver and is an example of coherent noise. It is easily recognizable because it travels at a speed of 330 m/s, the speed of sound in air.

Ground Roll/Rayleigh Wave/Scholte Wave/Surface Wave

A Rayleigh wave typically propagates along a free surface of a solid, but the elastic constants and density of air are very low compared to those of rocks so the surface of the Earth is approximately a free surface. Low velocity, low frequency and high amplitude Rayleigh waves are frequently present on a seismic record and can obscure signal, degrading overall data quality. They are known within the industry as 'Ground Roll' and are an example of coherent noise that can be attenuated with a carefully designed seismic survey. The Scholte wave is similar to ground roll but occurs at the sea-floor (fluid/solid interface) and it can possibly obscure and mask deep reflections in marine seismic records. The velocity of these waves varies with wavelength, so they are said to be dispersive and the shape of the wavetrain varies with distance.

Refraction/Head Wave/Conical Wave

A head wave refracts at an interface, travelling along it, within the lower medium and produces oscillatory motion parallel to the interface. This motion causes a disturbance in the upper medium that is detected on the surface. The same phenomenon is utilised in seismic refraction.

Multiple Reflection

An event on the seismic record that has incurred more than one reflection is called a *multiple*. Multiples can be either short-path (peg-leg) or long-path, depending upon whether they interfere with primary reflections or not.

Multiples from the bottom of a body of water (the interface of the base of water and the rock or sediment beneath it) and the air-water interface are common in marine seismic data, and are suppressed by seismic processing.

Cultural Noise

Cultural noise includes noise from weather effects, planes, helicopters, electrical pylons, and ships (in the case of marine surveys), all of which can be detected by the receivers.

Applications

Reflection seismology is used extensively in a number of fields and its applications can be categorised into three groups, each defined by their depth of investigation:

- Near-surface applications: An application that aims to understand geology at depths of up to approximately 1 km, typically used for engineering and environmental surveys, as well as coal and mineral exploration. A more recently developed application for seismic reflection is for geothermal energy surveys, although the depth of investigation can be up to 2 km deep in this case.

- Hydrocarbon exploration: Used by the hydrocarbon industry to provide a high resolution map of acoustic impedance contrasts at depths of up to 10 km within the subsurface. This can be combined with seismic attribute analysis and other exploration geophysics tools and used to help geologists build a geological model of the area of interest.

- Mineral exploration: The traditional approach to near-surface (<300 m) mineral exploration has been to employ geological mapping, geochemical analysis and the use of aerial and ground-based potential field methods, in particular for greenfield exploration, in the recent decades reflection seismic has become a valid method for exploration in hard-rock environments.

- Crustal studies: Investigation into the structure and origin of the Earth's crust, through to the Moho discontinuity and beyond, at depths of up to 100 km.

A method similar to reflection seismology which uses electromagnetic instead of elastic waves, and has a smaller depth of penetration, is known as Ground-penetrating radar or GPR.

Hydrocarbon Exploration

Reflection seismology, more commonly referred to as "seismic reflection" or abbreviated to "seismic" within the hydrocarbon industry, is used by petroleum geologists and geophysicists to map and interpret potential petroleum reservoirs. The size and scale of seismic surveys has increased alongside the significant concurrent increases in computer power during the last 25 years. This has led the seismic industry from laboriously – and therefore rarely – acquiring small 3D surveys in the 1980s to now routinely acquiring large-scale high resolution 3D surveys. The goals and basic principles have remained the same, but the methods have slightly changed over the years.

The primary environments for seismic exploration are land, the transition zone and marine:

Land: The land environment covers almost every type of terrain that exists on Earth, each bringing its own logistical problems. Examples of this environment are jungle, desert, arctic tundra, forest, urban settings, mountain regions and savannah.

Transition Zone: The transition zone is considered to be the area where the land meets the sea, presenting unique challenges because the water is too shallow for large seismic vessels but too deep for the use of traditional methods of acquisition on land. Examples of this environment are river deltas, swamps and marshes, coral reefs, beach tidal areas and the surf zone. Transition zone seismic crews will often work on land, in the transition zone and in the shallow water marine environment on a single project in order to obtain a complete map of the subsurface.

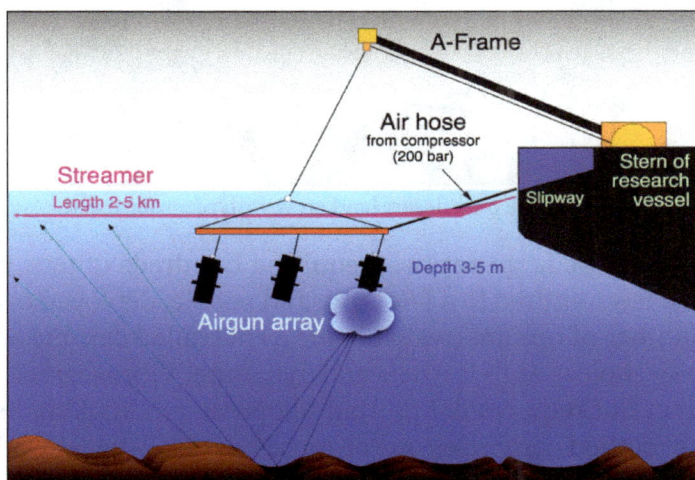

Diagram of equipment used for marine seismic surveys.

Marine: The marine zone is either in shallow water areas (water depths of less than 30 to 40 metres would normally be considered shallow water areas for 3D marine seismic operations) or in the deep water areas normally associated with the seas and oceans (such as the Gulf of Mexico).

Seismic surveys are typically designed by National oil companies and International oil companies who hire service companies such as CGG, Petroleum Geo-Services and WesternGeco to acquire them. Another company is then hired to process the data, although this can often be the same company that acquired the survey. Finally the finished seismic volume is delivered to the oil company so that it can be geologically interpreted.

Land Survey Acquisition

Desert land seismic camp.

Receiver line on a desert land crew with recorder truck.

Land seismic surveys tend to be large entities, requiring hundreds of tons of equipment and employing anywhere from a few hundred to a few thousand people, deployed over vast areas for many months. There are a number of options available for a controlled seismic source in a land survey and particularly common choices are Vibroseis and dynamite. Vibroseis is a non-impulsive source that is cheap and efficient but requires flat ground to operate on, making its use more difficult in undeveloped areas. The method comprises one or more heavy, all-terrain vehicles lowering a steel plate onto the ground, which is then vibrated with a specific frequency distribution and amplitude. It produces a low energy density, allowing it to be used in cities and other built-up areas where dynamite would cause significant damage, though the large weight attached to a Vibroseis truck can cause its own environmental damage. Dynamite is an impulsive source that is regarded as the ideal geophysical source due to it producing an almost perfect impulse function but it has obvious environmental drawbacks. For a long time, it was the only seismic source available until weight dropping was introduced around 1954, allowing geophysicists to make a trade-off between image quality and environmental damage. Compared to Vibroseis, dynamite is also operationally inefficient because each source point needs to be drilled and the dynamite placed in the hole.

A land seismic survey requires substantial logistical support. In addition to the day-to-day seismic operation itself, there must also be support for the main camp (for catering, waste management and laundry etc.), smaller camps (for example where the distance is too far to drive back to the main camp with vibrator trucks), vehicle and equipment maintenance, medical personnel and security.

Unlike in marine seismic surveys, land geometries are not limited to narrow paths of acquisition, meaning that a wide range of offsets and azimuths is usually acquired and the largest challenge is increasing the rate of acquisition. The rate of production is obviously controlled by how fast the source (Vibroseis in this case) can be fired and then move on to the next source location. Attempts have been made to use multiple seismic sources at the same time in order to increase survey efficiency and a successful example of this technique is Independent Simultaneous Sweeping (ISS).

Marine Survey Acquisition

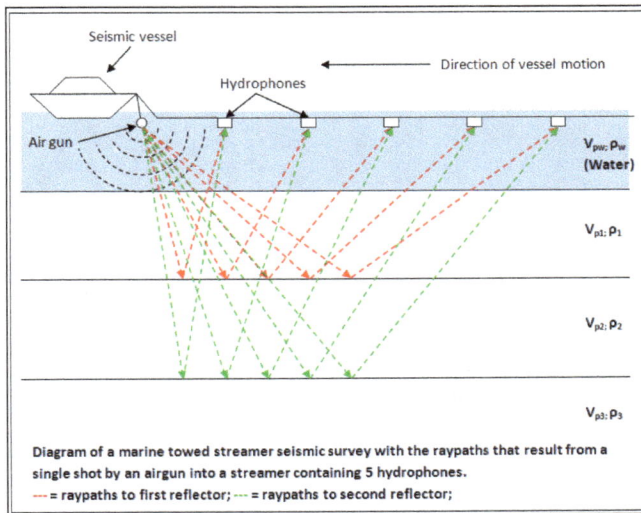

Diagram of a marine towed streamer seismic survey with the raypaths that result from a single shot by an airgun into a streamer containing 5 hydrophones.
--- = raypaths to first reflector; --- = raypaths to second reflector;

Marine seismic survey using a towed streamer.

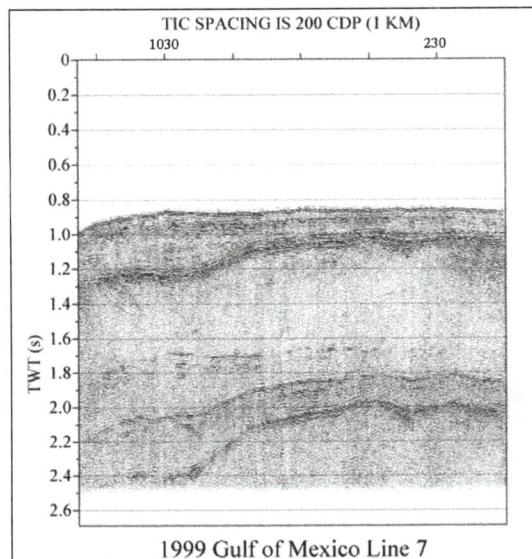

Seismic data collected by the USGS in the Gulf of Mexico.

Traditional marine seismic surveys are conducted using specially-equipped vessels that tow one or more cables containing a series of hydrophones at constant intervals. The cables are known as streamers, with 2D surveys using only 1 streamer and 3D surveys employing up to 12 or more (though 6 or 8 is more common). The streamers are deployed just beneath the surface of the water and are at a set distance away from the vessel. The seismic source, usually an airgun or an array of airguns but other sources are available, is also deployed beneath the water surface and is located between the vessel and the first receiver. Two identical sources are often used to achieve a faster rate of shooting. Marine seismic surveys generate a significant quantity of data, each streamer can be up to 6 or even 8 km long, containing hundreds of channels and the seismic source is typically fired every 15 or 20 seconds.

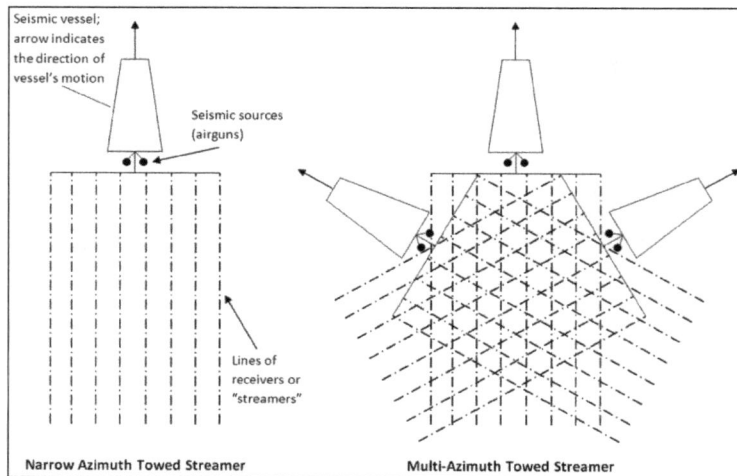

Plan view of NATS and MAZ surveys.

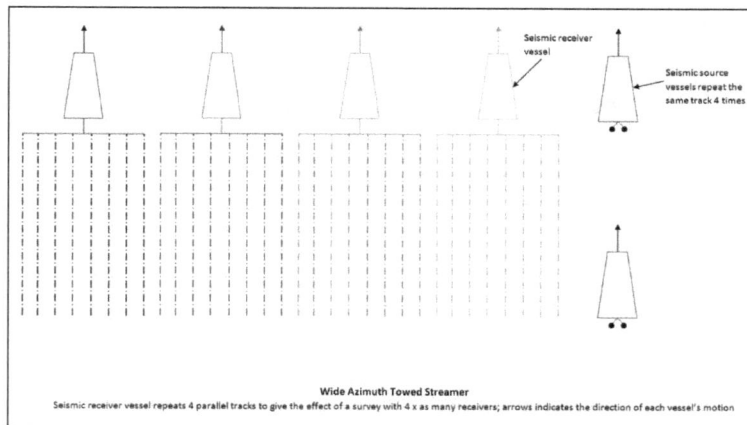

Plan view of a WATS/WAZ survey.

A seismic vessel with 2 sources and towing a single streamer is known as a Narrow-Azimuth Towed Streamer (or NAZ or NATS). By the early 2000s, it was accepted that this type of acquisition was useful for initial exploration but inadequate for development and production, in which wells had to be accurately positioned. This led to the development of the Multi-Azimuth Towed Streamer (MAZ) which tried to break the limitations of the linear acquisition pattern of a NATS survey by acquiring a combination of NATS surveys at different azimuths. This successfully delivered increased illumination of the subsurface and a better signal to noise ratio.

The seismic properties of salt poses an additional problem for marine seismic surveys, it attenuates seismic waves and its structure contains overhangs that are difficult to image. This led to another variation on the NATS survey type, the wide-azimuth towed streamer (or WAZ or WATS) and was first tested on the Mad Dog field in 2004. This type of survey involved 1 vessel solely towing a set of 8 streamers and 2 separate vessels towing seismic sources that were located at the start and end of the last receiver line. This configuration was "tiled" 4 times, with the receiver vessel moving further away from the source vessels each time and eventually creating the effect of a survey with 4 times the number of streamers. The end result was a seismic dataset with a larger range of wider azimuths, delivering a breakthrough in seismic imaging. These are now the three common types of marine towed streamer seismic surveys.

Marine survey acquisition is not just limited to seismic vessels; it is also possible to lay cables of geophones and hydrophones on the sea bed in a similar way to how cables are used in a land seismic survey, and use a separate source vessel. This method was originally developed out of operational necessity in order to enable seismic surveys to be conducted in areas with obstructions, such as production platforms, without having the compromise the resultant image quality. Ocean bottom cables (OBC) are also extensively used in other areas that a seismic vessel cannot be used, for example in shallow marine (water depth <300m) and transition zone environments, and can be deployed by remotely operated underwater vehicles (ROVs) in deep water when repeatability is valued. Conventional OBC surveys use dual-component receivers, combining a pressure sensor (hydrophone) and a vertical particle velocity sensor (vertical geophone), but more recent developments have expanded the method to use four-component sensors i.e. a hydrophone and three orthogonal geophones. Four-component sensors have the advantage of being able to also record shear waves, which do not travel through water but can still contain valuable information.

In addition to the operational advantages, OBC also has geophysical advantages over a conventional NATS survey that arise from the increased fold and wider range of azimuths associated with the survey geometry. However, much like a land survey, the wider azimuths and increased fold come at a cost and the ability for large-scale OBC surveys is severely limited.

In 2005, ocean bottom nodes (OBN) – an extension of the OBC method that uses battery-powered cableless receivers placed in deep water – was first trialled over the Atlantis Oil Field in a partnership between BP and Fairfield Industries. The placement of these nodes can be more flexible than the cables in OBC and they are easier to store and deploy due to their smaller size and lower weight.

Time Lapse Acquisition

Time lapse or 4D surveys are 3D seismic surveys repeated after a period of time. The 4D refers to the fourth dimension which in this case is time. Time lapse surveys are acquired in order to observe reservoir changes during production and identify areas where there are barriers to flow that may not be detectable in conventional seismic. Time lapse surveys consist out of a baseline survey and a monitor or repeat survey, acquired after the field was under production. Most of these surveys have been repeated NATS surveys as they are cheaper to acquire and most fields historically already had a NATS baseline survey. Some of these surveys are collected using ocean-bottom cables because the cables can be accurately placed in their previous location after being removed. Better repetition of the exact source and receiver location leads to improved repeatability and better signal to noise ratios. A number of 4D surveys have also been set up over fields in which ocean bottom cables have been purchased and permanently deployed. This method can be known as life of field seismic (LoFS) or permanent reservoir monitoring (PRM).

OBN has proven to be another very good way to accurately repeat a seismic acquisition. The world's first 4D survey using nodes was acquired over the Atlantis Oil Field in 2009, with the nodes being placed by a ROV in a water depth of 1300–2200 m to within a few meters of where they were previously placed in 2005.

Seismic Data Processing

There are three main processes in seismic data processing: deconvolution, common-midpoint (CMP) stacking and migration.

Deconvolution is a process that tries to extract the reflectivity series of the Earth, under the assumption that a seismic trace is just the reflectivity series of the Earth convolved with distorting filters. This process improves temporal resolution by collapsing the seismic wavelet, but it is nonunique unless further information is available such as well logs, or further assumptions are made. Deconvolution operations can be cascaded, with each individual deconvolution designed to remove a particular type of distortion.

CMP stacking is a robust process that uses the fact that a particular location in the subsurface will have been sampled numerous times and at different offsets. This allows a geophysicist to construct a group of traces with a range of offsets that all sample the same subsurface location, known as a Common Midpoint Gather. The average amplitude is then calculated along a time sample, resulting in significantly lowering the random noise but also losing all valuable information about the relationship between seismic amplitude and offset. Less significant processes that are applied shortly before the CMP stack are Normal moveout correction and statics correction. Unlike marine seismic data, land seismic data has to be corrected for the elevation differences between the shot and receiver locations. This correction is in the form of a vertical time shift to a flat datum and is known as a statics correction, but will need further correcting later in the processing sequence because the velocity of the near-surface is not accurately known. This further correction is known as a residual statics correction.

Seismic migration is the process by which seismic events are geometrically re-located in either space or time to the location the event occurred in the subsurface rather than the location that it was recorded at the surface, thereby creating a more accurate image of the subsurface.

Seismic Interpretation

The goal of seismic interpretation is to obtain a coherent geological story from the map of processed seismic reflections. At its most simple level, seismic interpretation involves tracing and correlating along continuous reflectors throughout the 2D or 3D dataset and using these as the basis for the geological interpretation. The aim of this is to produce structural maps that reflect the spatial variation in depth of certain geological layers. Using these maps hydrocarbon traps can be identified and models of the subsurface can be created that allow volume calculations to be made. However, a seismic dataset rarely gives a picture clear enough to do this. This is mainly because of the vertical and horizontal seismic resolution but often noise and processing difficulties also result in a lower quality picture. Due to this, there is always a degree of uncertainty in a seismic interpretation and a particular dataset could have more than one solution that fits the data. In such a case, more data will be needed to constrain the solution, for example in the form of further seismic acquisition, borehole logging or gravity and magnetic survey data. Similarly to the mentality of a seismic processor, a seismic interpreter is generally encouraged to be optimistic in order encourage further work rather than the abandonment of the survey area. Seismic interpretation is completed by both geologists and geophysicists, with most seismic interpreters having an understanding of both fields.

In hydrocarbon exploration, the features that the interpreter is particularly trying to delineate are the parts that make up a petroleum reservoir – the source rock, the reservoir rock, the seal and trap.

Seismic Attribute Analysis

Seismic attribute analysis involves extracting or deriving a quantity from seismic data that can be analysed in order to enhance information that might be more subtle in a traditional seismic image, leading to a better geological or geophysical interpretation of the data. Examples of attributes that can be analysed include mean amplitude, which can lead to the delineation of bright spots and dim spots, coherency and amplitude versus offset. Attributes that can show the presence of hydrocarbons are called direct hydrocarbon indicators.

Crustal Studies

The use of reflection seismology in studies of tectonics and the Earth's crust was pioneered in the 1970s by groups such as the Consortium for Continental Reflection Profiling (COCORP), who inspired deep seismic exploration in other countries such as BIRPS in Great Britain and ECORS in France. The British Institutions Reflection Profiling Syndicate (BIRPS) was started up as a result of oil hydrocarbon exploration in the North Sea. It became clear that there was a lack of understanding of the tectonic processes that had formed the geological structures and sedimentary basins which were being explored. The effort produced some significant results and showed that it is possible to profile features such as thrust faults that penetrate through the crust to the upper mantle with marine seismic surveys.

Swell Filter

The term swell filter in high resolution seismics (reflection seismology) or sub bottom profiling refers to the static correction that restores the coherence of a high resolution seismic profile. The coherence of the image got lost because of the relative movement (a function of the wavelength of the signal and the swell) of the source and receiver during the recording. In normal seismic recordings, the term swell filter refers to filtering the acoustic noise, created by waves, out of the seismic recording.

High Resolution Seismics and Subbottom Profiling

Very high resolution seismics (reflection seismology) is applied to make detailed acoustic profiles of a sea floor or lake floor. Acoustic sources used include: air gun, water gun, sleeve gun, sparker and boomer. The frequency used in this field ranges from 100 Hz to 10.000 Hz.

Subbottom profiling is a technique used in acoustical oceanography. Sonars or echo sounder are typical sources for very high resolution imaging. Also used in this field are parametric echosounders. The main difference with the high resolution seismics is the type of source used.

Problematic Swell

When the movement of the platform and equipment due to swell is too heavy during a seismic recording, the coherence of the image gets lost. The reason why it gets lost is because the movement of source and receiver is larger than the wavelength of the signal produced by the source. As such the seismic reflectors do not line up.

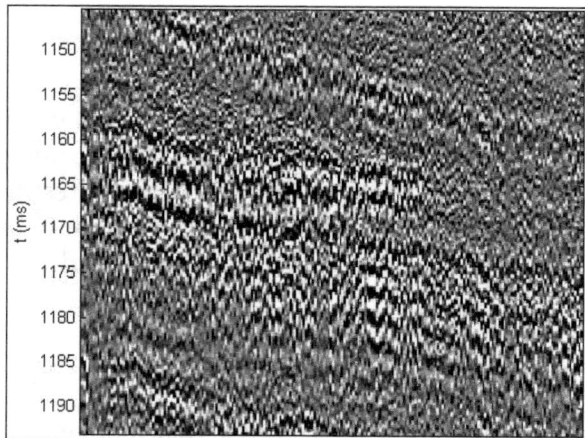

A detail of a seismic profile recorded
during heavy weather.

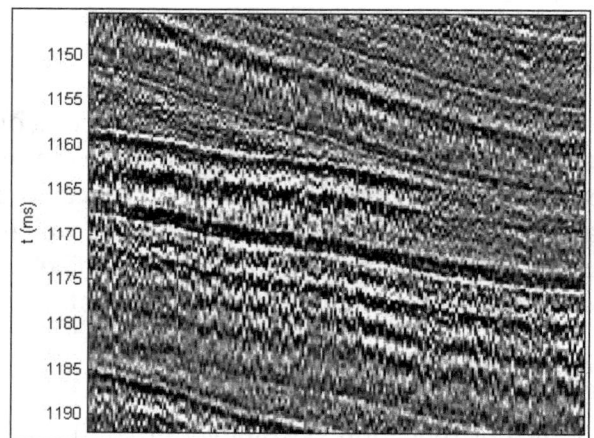

The same profile swell filtered.

SEISMIC REFRACTION

Seismic refraction is a geophysical principle governed by Snell's Law. Used in the fields of engineering geology, geotechnical engineering and exploration geophysics, seismic refraction traverses (seismic lines) are performed using a seismograph(s) and/or geophone(s), in an array and an energy source. The seismic refraction method utilizes the refraction of seismic waves on geologic layers and rock/soil units in order to characterize the subsurface geologic conditions and geologic structure.

The methods depend on the fact that seismic waves have differing velocities in different types of soil (or rock): In addition, the waves are refracted when they cross the boundary between different types (or conditions) of soil or rock. The methods enable the general soil types and the approximate depth to strata boundaries, or to bedrock, to be determined.

P-wave Refraction

P-wave refraction evaluates the compression wave generated by the seismic source located at a known distance from the array. The wave is generated by vertically striking a striker plate with a sledgehammer, shooting a seismic shotgun into the ground, or detonating an explosive charge in the ground. Since the compression wave is the fastest of the seismic waves, it is sometimes referred to as the primary wave and is usually more-readily identifiable within the seismic recording as compared to the other seismic waves.

S-wave Refraction

S-wave refraction evaluates the shear wave generated by the seismic source located at a known distance from the array. The wave is generated by horizontally striking an object on the ground surface to induce the shear wave. Since the shear wave is the second fastest wave, it is sometimes referred to as the secondary wave. When compared to the compression wave, the shear wave is approximately one-half (but may vary significantly from this estimate) the velocity depending on the medium.

Two Horizontal Layers

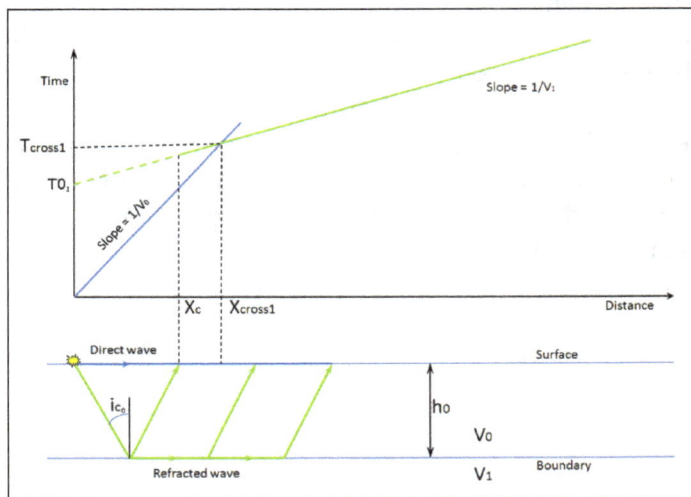

i_{co} - Critical angle.

V_0 - Velocity of the first layer.

V_1 - Velocity of the second layer.

h_0 - Thickness of the first layer.

To_1 - Intercept.

$$i_{c_0} = asin\left(\frac{V_0}{V_1}\right)$$

$$T = \frac{2h_0 cos(i_{c_0})}{V_0} + \frac{X}{V_1} = T0_1 + \frac{X}{V_1}$$

$$h_0 = \frac{T0_1 V_0}{2cos(i_c)}$$

$$h_0 = \frac{X_{cross_1}}{2}\sqrt{\frac{V_1 - V_0}{V_1 + V_0}}$$

Several Horizontal Layers

$$h_n = \frac{V_n}{cos(i_n)}\left(\frac{T0_{n+1}}{2} - \sum_{j=0}^{n-1} h_j \sqrt{\frac{1}{V_j^2} - \frac{1}{V_{j+1}^2}}\right)$$

ISOSEISMAL MAP

In seismology, an isoseismal map is used to show lines of equal felt seismic intensity, generally measured on the Modified Mercalli scale. Such maps help to identify earthquake epicenters, particularly where no instrumental records exist, such as for historical earthquakes. They also contain important information on ground conditions at particular locations, the underlying geology, radiation pattern of the seismic waves and the response of different types of buildings. They form an important part of the macroseismic approach, i.e. that part of seismology dealing with non-instrumental data. The shape and size of the isoseismal regions can be used to help determine the magnitude, focal depth and focal mechanism of an earthquake.

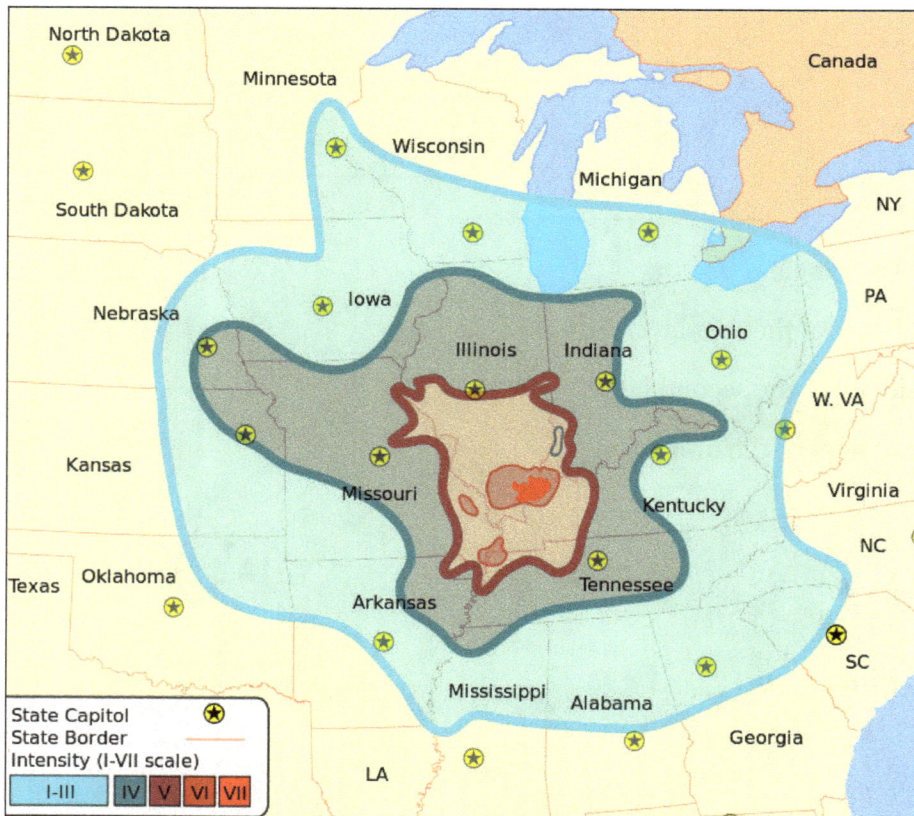

Isoseismal map for the 1968 Illinois earthquake.

Methodology

Firstly, observations of the felt intensity need to be obtained for all areas affected by the tremor. In the case of recent earthquakes, news reports are augmented by sending out questionnaires or by collecting information online about the intensity of the shaking. For a historical earthquake the procedure is much the same, except that it requires searching through contemporary accounts in newspapers, letters, diaries, etc. Once the information has been assembled and intensities assigned at the location of the individual observations, these are plotted on a map. Isoseismal lines are then drawn to link together areas of equal shaking. Because of local variations in the ground conditions, isoseismals will generally separate zones of broadly similar felt intensity, while containing areas of both higher and lower degrees of shaking. To make the isoseismals less subjective, attempts have been made to use computer-based methods of contouring such as kriging, rather than relying on visual interpolation.

Use

Locating the Epicenter

In most earthquakes, the isoseismals define a single clear area of maximum intensity, which is known as the epicentral area. In some earthquakes, there is more than one maximum because of the effect of ground conditions or complexities in the rupture propagation, and other information is therefore required to identify the area that contains the epicenter.

Measuring the Magnitude

The magnitude of an earthquakes can be roughly estimated by measuring the area affected by intensity level III or above in square kilometres and taking the logarithm. A more accurate estimate relies on the development of regional calibration functions derived using many isoseismal radii. Such approaches allow magnitudes to be estimated for historical earthquakes.

Estimating the Focal Depth

The depth to the hypocenter can be estimated by comparing the sizes of different isoseismal areas. In shallow earthquakes the lines are close together while in deep events the lines are spread further apart.

Confirming the Focal Mechanism

Focal mechanisms are routinely calculated using teleseismic data but an ambiguity remains as there are always two potential fault planes. The shape of the areas of highest intensity are generally elongate along the direction of the active fault plane.

Testing Seismic Hazard Assessments

Because of the relatively long history of macroseismic intensity observations (sometimes stretching back many centuries in some regions) isoseismal maps can be used to test seismic hazard assessments by comparing the expected temporal frequency of different levels of intensity assuming an assessment is true and the observed rate of exceedance.

LINEAR SEISMIC INVERSION

Inverse modeling is a mathematical technique where the objective is to determine the physical properties of the subsurface of an earth region that has produced a given seismogram. Cooke and Schneider defined it as calculation of the earth's structure and physical parameters from some set of observed seismic data. The underlying assumption in this method is that the collected seismic data are from an earth structure that matches the cross-section computed from the inversion algorithm. Some common earth properties that are inverted for include acoustic velocity, formation and fluid densities, acoustic impedance, Poisson's ratio, formation compressibility, shear rigidity, porosity, and fluid saturation.

The method has long been useful for geophysicists and can be categorized into two broad types: Deterministic and stochastic inversion. Deterministic inversion methods are based on comparison of the output from an earth model with the observed field data and continuously updating the earth model parameters to minimize a function, which is usually some form of difference between model output and field observation. As such, this method of inversion to which linear inversion falls under is posed as a minimization problem and the accepted earth model is the set of model parameters that minimizes the objective function in producing a numerical seismogram which best compares with collected field seismic data.

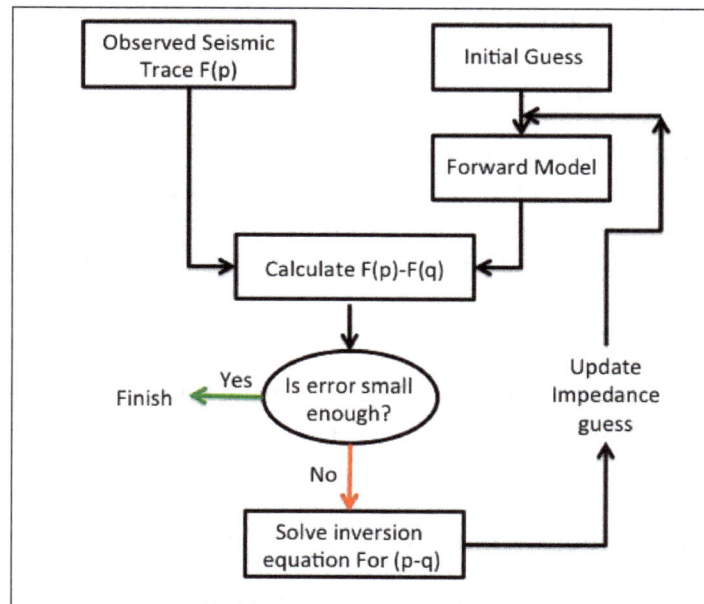

Linear Seismic Inversion Flow Chart.

On the other hand, stochastic inversion methods are used to generate constrained models as used in reservoir flow simulation, using geostatistical tools like kriging. As opposed to deterministic inversion methods which produce a single set of model parameters, stochastic methods generate a suite of alternate earth model parameters which all obey the model constraint. However, the two methods are related as the results of deterministic models is the average of all the possible non-unique solutions of stochastic methods. Since seismic linear inversion is a deterministic inversion method, the stochastic method will not be discussed beyond this point.

Linear Inversion

The deterministic nature of linear inversion requires a functional relationship which models, in terms of the earth model parameters, the seismic variable to be inverted. This functional relationship is some mathematical model derived from the fundamental laws of physics and is more often called a forward model. The aim of the technique is to minimize a function which is dependent on the difference between the convolution of the forward model with a source wavelet and the field collected seismic trace. As in the field of optimization, this function to be minimized is called the objective function and in convectional inverse modeling, is simply the difference between the convolved forward model and the seismic trace. Different types of variables can be inverted for but for clarity, these variables will be referred to as the impedance series of the earth model. In the following subsections we will describe in more detail, in the context of linear inversion as a minimization problem, the different components that are necessary to invert seismic data.

Forward Model

The centerpiece of seismic linear inversion is the forward model which models the generation of the experimental data collected. It provides a functional (computational) relationship between the model parameters and calculated values for the observed traces. Depending on the seismic data collected, this model may vary from the classical wave equations for predicting particle displacement or fluid pressure for sound wave propagation through rock or fluids, to some variants of these classical equations. For example, the forward model in Tarantola is the wave equation for pressure variation in a liquid media during seismic wave propagation while by assuming constant velocity layers with plane interfaces, Kanasewich and Chiu used the brachistotrone model of John Bernoulli for travel time of a ray along a path. In Cooke and Schneider, the model is a synthetic trace generation algorithm expressed as in equation where R(t) is generated in the Z-domain by recursive formula. In whatever form the forward model appears, it is important that it not only predicts the collected field data, but also models how the data is generated. Thus, the forward model by Cooke and Schneider can only be used to invert CMP data since the model invariably assumes no spreading loss by mimicking the response of a laterally homogeneous earth to a plane-wave source.

$$t = \sum_{i=1}^{n} \frac{[(x_i - x_{i-1})^2 + (y_i - y_{i-1})^2 + (z_i - z_{i-1})^2]^{\frac{1}{2}}}{v_i}$$

where t is ray travel time, x, y, z are depth coordinates and vi is the constant velocity between interfaces i − 1 and i.

$$\left[\frac{1}{K(\vec{r})} \frac{\partial^2}{\partial t^2} - \nabla \cdot \left(\frac{1}{\rho(\vec{r})} \nabla \right) \right] U(\vec{r}, t) = s(\vec{r}, t)$$

where $K(\vec{r})$ represent bulk modulus, $\rho(\vec{r})$ density, $s(\vec{r},t)$ the source of acoustic waves, and $U(\vec{r},t)$ the pressure variation.

$$s(t) = w(t) * R(t)$$

where $s(t)$ = synthetic trace, $w(t)$ = source wavelet, and $R(t)$ = reflectivity function.

Objective Function

An important numerical process in inverse modeling is to minimize the objective function, which is a function defined in terms of the difference between the collected field seismic data and the numerically computed seismic data. Classical objective functions include the sum of squared deviations between experimental and numerical data, as in the least squares methods, the sum of the magnitude of the difference between field and numerical data, or some variant of these definitions. Irrespective of the definition used, numerical solution of the inverse problem is obtained as earth model that minimize the objective function.

In addition to the objective function, other constraints like known model parameters and known layer interfaces in some regions of the earth are also incorporated in the inverse modeling procedure. These constraints, according to Francis 2006, help to reduce non-uniqueness of the inversion solution by providing a priori information that is not contained in the inverted data while Cooke and Schneider reports their useful in controlling noise and when working in a geophysically well-known area.

Mathematical Analysis of Generalized Linear Inversion Procedure

The objective of mathematical analysis of inverse modeling is to cast the generalized linear inverse problem into a simple matrix algebra by considering all the components viz; forward model, objective function etc. Generally, the numerically generated seismic data are non-linear functions of the earth model parameters. To remove the non-linearity and create a platform for application of linear algebra concepts, the forward model is linearized by expansion using a Taylor series as carried out below.

Consider a set of m seismic field observations F_j, for $j = 1, \ldots, m$ and a set of n earth model parameters p_i to be inverted for, for $i = 1, \ldots, n$. The field observations can be represented in either $\vec{F}(\vec{p})$ or $F_j(p_i)$, where $\vec{p}$ and $\vec{F}(\vec{p})$ are vectorial representations of model parameters and the field observations as a function of earth parameters. Similarly, for q_i representing guesses of model parameters, $\vec{F}(\vec{q})$ is the vector of numerical computed seismic data using the forward model. Taylor's series expansion of $\vec{F}(\vec{p})$ about $\vec{q}$ is given below:

$$\vec{F}(\vec{p}) = \vec{F}(\vec{q}) + (\vec{p} - \vec{q})\frac{\partial \vec{F}(\vec{q})}{\partial \vec{p}} + (\vec{p} - \vec{q})^2 \frac{\partial^2 \vec{F}(\vec{q})}{\partial \vec{p}^2} + O(\vec{p} - \vec{q})^3$$

On linearization by dropping the non-linear terms (terms with (p⃗ – ⃗q) of order 2 and above), the equation becomes:

$$\vec{F}(\vec{p}) - \vec{F}(\vec{q}) = (\vec{p} - \vec{q})\frac{\partial \vec{F}(\vec{q})}{\partial \vec{p}}$$

Considering that $\vec{F}$ has m components and $\vec{p}$ and $\vec{q}$ have n components, the discrete form of Eqn. results in a system of linear equations in n variables whose matrix form is shown below:

$$\Delta \vec{F} = \mathbf{A}\ddot{A}\vec{p}$$

$$\Delta \vec{F} = \begin{bmatrix} F_1(\vec{p}) - F_1(\vec{q}) \\ \vdots \\ F_m(\vec{p}) - F_m(\vec{q}) \end{bmatrix}$$

$$\Delta \vec{p} = \vec{p} - \vec{q} = \begin{bmatrix} p_1 - q_1 \\ \vdots \\ p_n - q_n \end{bmatrix}$$

$$\mathbf{A} = \begin{bmatrix} \dfrac{\partial F_1(\vec{q})}{\partial p_1} & \dfrac{\partial F_1(\vec{q})}{\partial p_2} & \cdots & \dfrac{\partial F_1(\vec{q})}{\partial p_n} \\[2mm] \dfrac{\partial F_2(\vec{q})}{\partial p_1} & \cdots & \dfrac{\partial F_2(\vec{q})}{\partial p_{n-1}} & \dfrac{\partial F_2(\vec{q})}{\partial p_n} \\[2mm] \vdots & \dfrac{\partial F_j(\vec{q})}{\partial p_i} & \vdots & \vdots \\[2mm] \dfrac{\partial F_m(\vec{q})}{\partial p_1} & \dfrac{\partial F_m(\vec{q})}{\partial p_2} & \cdots & \dfrac{\partial F_m(\vec{q})}{\partial p_n} \end{bmatrix}$$

$\Delta \vec{F}$ is called the difference vector in Cooke and Schneider. It has a size of $m \times 1$ and its components are the difference between the observed trace and the numerically computed seismic data. $\Delta \vec{p}$ is the corrector vector of size $\Delta \vec{p}$, while $\mathbf{A}$ is called the sensitivity matrix. It has a size of $m \times n$ and its comments are such that each column is the partial derivative of a component of the forward function with respect to one of the unknown earth model parameters. Similarly, each row is the partial derivative of a component of the numerically computed seismic trace with respect to all unknown model parameters.

Solution Algorithm

$\vec{F}(\vec{q})$ is computed from the forward model, while $\vec{F}(\vec{p})$ is the experimental data. Thus, $\Delta \vec{F}$ is a known quality. On the other hand, $\Delta \vec{p}$ is unknown and is obtained by solution of equation. This equation is theoretically solvable only when $\mathbf{A}$ is invertible, that is, if it is a square matrix so that the number of observations m is equal to the number n of unknown earth parameters. If this is the case, the unknown corrector vector $\Delta \vec{p}$, is solved for as shown below, using any of the classical direct or iterative solvers for solution of a set of linear equations:

$$\Delta \vec{p} = \mathbf{A}^{-1} \Delta \vec{F}$$

In most seismic inversion applications, there are more observations than the number of earth parameters to be inverted for, i.e. $m > n$, leading to a system of equations that is mathematically over-determined. As a result, the equation is not theoretically solvable and an exact solution is not obtainable. An estimate of the corrector vector is obtained using the least squares procedure to find the corrector vector $\Delta \vec{p}$ that minimizes $\vec{e}^T \vec{e}$, which is the sum of the squares of the error, $\vec{e}$.

The error $\vec{e}$ is given by:

$$\vec{e} = \Delta \vec{F} - \mathbf{A} \Delta \vec{p}$$

In the least squares procedure, the corrector vector that minimizes $\vec{e}^T\vec{e}$ is obtained as below:

$$\mathbf{A}\Delta\vec{p} = \Delta\vec{F}$$
$$\mathbf{A}^T\mathbf{A}\Delta\vec{p} = \mathbf{A}^T\Delta\vec{F}$$

Thus,

$$\Delta\vec{p} = (\mathbf{A}^T\mathbf{A})^{-1}\mathbf{A}^T\Delta\vec{F}$$

From the above discussions, the objective function is defined as either the L_1 or L_1 norm of $\ddot{\mathbf{A}}\vec{p}$ given by $\sum_{j=0}^{n}|\Delta p_j|$ or $\sum_{j=0}^{n}|\Delta p_j|^2$ or of $\Delta\vec{F}$ given by $\sum_{i=0}^{m}|\Delta F_i|$ or $\sum_{i=0}^{m}|\Delta F_i|^2$.

The generalized procedure for inverting any experimental seismic data for $m = n$ or $m > n$, using the mathematical theory for inverse modelling, are described as follows.

An initial guess of the model impedance is provided to initiate the inversion process. The forward model uses this initial guess to compute a synthetic seismic data which is subtracted from the observed seismic data to calculate the difference vector.

- An initial guess of the model impedance $\vec{q}$ is provided to initiate the inversion process.

- A synthetic seismic data $\vec{F}(\vec{q})$ is computed by the forward model, using the model impedance above.

- The difference vector $\vec{F}(\vec{p}) - \vec{F}(\vec{q})$ is computed as the difference between experimental and synthetic seismic data.

- The sensitivity matrix A is computed at this value of the impedance profile.

- Using A and the difference vector from 3 above, the corrector vector $\Delta\vec{p}$ is calculated. A new impedance profile is obtained as:

    ○ $\vec{p} = \vec{q} + \Delta\vec{p}$

- The L_1 or L_1 norm of the computed corrector vector is compared with a provided tolerance value. If the computed norm is less than the tolerance, the numerical procedure is concluded and the inverted impedance profile for the earth region is given by $\vec{p}$ from equation. On the other hand, if the norm is greater than the tolerance, iterations through steps 2-6 are repeated but with an updated impedance profile as computed from equation. According to Cooke and Schneider, use of the corrected guess from equation ($\vec{p} = \vec{q} + \Delta\vec{p}$) as the new initial guess during iteration reduces the error.

Parameterization of the Earth Model Space

Irrespective of the variable to be inverted for, the earth's impedance is a continuous function of depth (or time in seismic data) and for numerical linear inversion technique to be applicable for this continuous physical model, the continuous properties have to be discretized and/or sampled at discrete intervals along the depth of the earth model. Thus, the total depth over which model properties are to be determined is a necessary starting point for the discretization. Commonly, as shown in figure, these properties are sampled at close discrete intervals over this depth to ensure high resolution of impedance variation along the earth's depth. The impedance values inverted from the algorithm represents the average value in the discrete interval.

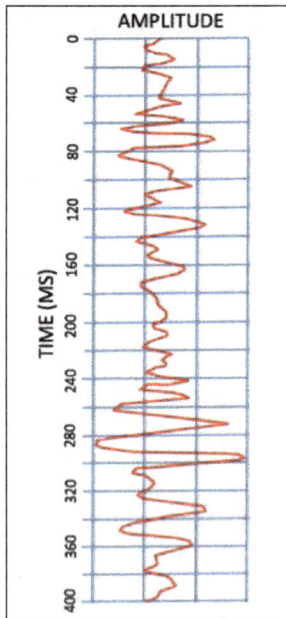

Amplitude Log.

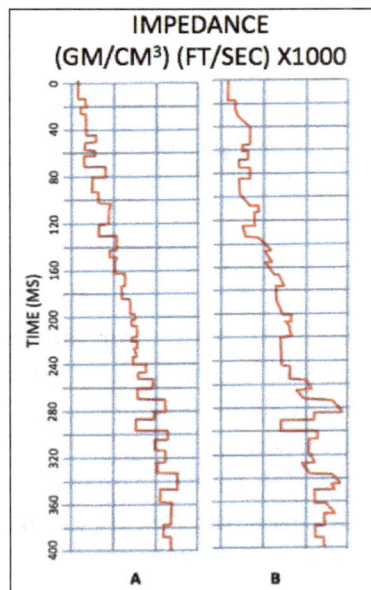

Impedance Logs Inverted From Amplitude.

Considering that inverse modeling problem is only theoretically solvable when the number of discrete intervals for sampling the properties is equal to the number of observation in the trace to be inverted, a high-resolution sampling will lead to a large matrix which will be very expensive to invert. Furthermore, the matrix may be singular for dependent equations, the inversion can be unstable in the presence of noise and the system may be under-constrained if parameters other than the primary variables inverted for, are desired. In relation to parameters desired, other than impedance, Cooke and Schneider gives them to include source wavelet and scale factor.

Finally, by treating constraints as known impedance values in some layers or discrete intervals, the number of unknown impedance values to be solved for are reduced, leading to greater accuracy in the results of the inversion algorithm.

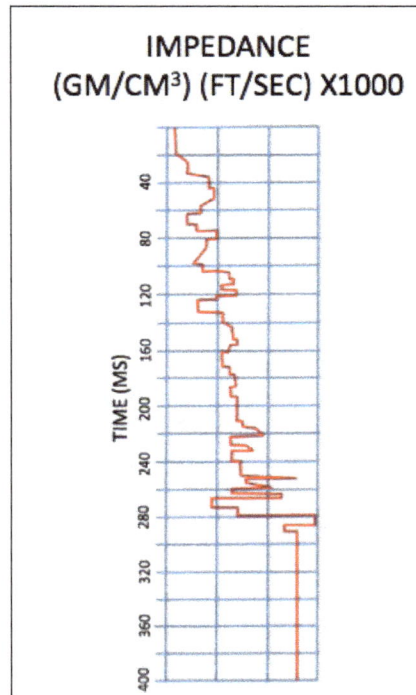

Impedance Well Log.

Inversion Examples

Temperature Inversion

We start with an example to invert for earth parameter values from temperature depth distribution in a given earth region. Although this example does not directly relate to seismic inversion since no traveling acoustic waves are involved, it nonetheless introduces practical application of the inversion technique in a manner easy to comprehend, before moving on to seismic applications. In this example, the temperature of the earth is measured at discrete locations in a well bore by placing temperature sensors in the target depths. By assuming a forward model of linear distribution of temperature with depth, two parameters are inverted for from the temperature depth measurements.

The forward model is given by:

$$\vec{F}(\vec{q}) = \vec{T} = a + bz$$

where $\vec{q} = [a, b]$. Thus, the dimension of $\vec{q}$ is 2 i.e. the number of parameters inverted for is 2.

The objective of this inversion algorithm is to find $\vec{p}$, which is the value of $[a,b]$ that minimizes the difference between the observed temperature distribution and those obtained using the forward model of equation. Considering the dimension of the forward model or the number of temperature observations to be n, the components of the forward model is written as:

$$T_1 = a + bz_1$$
$$T_2 = a + bz_2$$
$$\vdots$$
$$T_{n-1} = a + bz_{n-1}$$
$$T_n = a + bz_n$$

so that $\vec{F}(\vec{q}) = T$,

$$A = \begin{bmatrix} 1 & z_1 \\ 1 & z_2 \\ \vdots & \vdots \\ 1 & z_{n-1} \\ 1 & z_n \end{bmatrix}$$

We present results for the case of $n = 2$ for which the observed temperature values at depths were $T_1 = 19°C$ at $z = 2m$ and $T_2 = 22°C$ at z m. These experimental data were inverted to obtain earth parameter values of $a = 0.5$ and $b = 18°C$. For a more general case with large number of temperature observations, the final linear forward model obtained from using the inverted values of a and b.

Seismic Trace Inversion

This example shows inversion of a CMP seismic trace for earth model impedance (product of density and velocity) profile. The seismic trace inverted is the inverted impedance profile with the input initial impedance used for the inversion algorithm. Also recorded alongside the seismic trace is an impedance log of the earth region. The figures show good comparison between the recorded impedance log and the numerical inverted impedance from the seismic trace.

PASSIVE SEISMIC

Passive seismic is the detection of natural low frequency earth movements, usually with the purpose of discerning geological structure and locate underground oil, gas, or other resources. Usually the data listening is done in multiple measurement points that are separated by several hundred meters, over periods of several hours to several days, using portable seismometers. The conclusions about the geological structure are based on the spectral analysis or on the mathematical reconstruction of the propagation and possible sources of the observed seismic waves. If the latter is planned, data are usually acquired in multiple (in the ideal case - all) points simultaneously, using so called synchronized lines. Reliability of the time reverse modelling can be further increased using results of reflection seismology about the distribution of the sound speed in the underground volume.

The low frequency 3-direction seismometer, suitable for the passive seismic.

Passive seismic usually focuses on a low frequency signals (0 to 10 Hz) and is sometimes called the "low frequency" seismology. The seismometers record movements in all 3 possible directions independently (such devices also have other application areas like long-term measurement stations). When needed, data acquisition is also done under water, using waterproof devices that measure earth movements at the bottom of the sea. Geophones are almost never used due to their limited sensitivity.

The survey using this method is very different from the conventional survey that is usually based on reflection seismology. The conventional survey consists of numerous measurements that are spatially very close together and relatively short (often lasting in the order of minutes). The passive seismic survey has much less measurements but they are frequently recorded for days. Local time must be taken into consideration, picking intervals with less human induced noise. Even relatively distant earthquakes are visible in the recorded spectrograms and must also be excluded from analysis.

The similar method have been also applied in another planets. For instance, during Apollo missions, the Passive Seismic Experiment sensors were deployed that detected lunar "moonquakes" and provided information about the internal structure of the Moon.

Passive seismic is much less expensive than well drilling. It is also cheaper and more environmentally friendly than active seismic, which requires a strong source of the seismic waves (like an underground explosion) to predict the structure. In some cases it may be the only method for which land access is granted by the land owner. Despite the method having been successfully applied in many parts of the world, this approach is currently less reliable, as the scientific methods are still largely under development.

SEISMIC MAGNITUDE SCALES

Seismic magnitude scales are used to describe the overall strength or "size" of an earthquake. These are distinguished from seismic intensity scales that categorize the intensity or severity of ground shaking (quaking) caused by an earthquake at a given location. Magnitudes are usually determined from measurements of an earthquake's seismic waves as recorded on a seismogram. Magnitude scales vary on what aspect of the seismic waves are measured and how they are measured. Different magnitude scales are necessary because of differences in earthquakes, the information available, and the purposes for which the magnitudes are used.

Earthquake Magnitude and Ground-shaking Intensity

The Earth's crust is stressed by tectonic forces. When this stress becomes great enough to rupture the crust, or to overcome the friction that prevents one block of crust from slipping past another, energy is released, some of it in the form of various kinds of seismic waves that cause ground-shaking, or quaking.

Magnitude is an estimate of the relative "size" or strength of an earthquake, and thus its potential for causing ground-shaking. It is "approximately related to the released seismic energy."

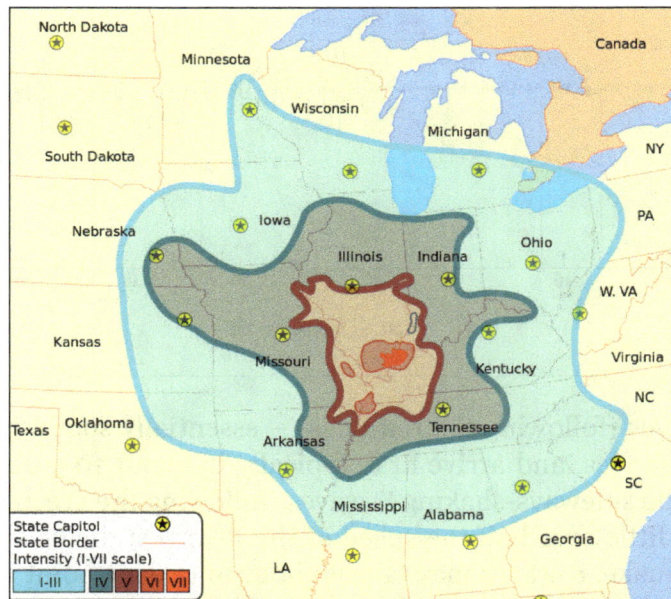

Isoseismal map. The irregular distribution of shaking arises from variations of geology or ground conditions.

Intensity refers to the strength or force of shaking at a given location, and can be related to the peak ground velocity. With an isoseismal map of the observed intensities an earthquake's magnitude can be estimated from both the maximum intensity observed (usually but not always near the epicenter), and from the extent of the area where the earthquake was felt.

The intensity of local ground-shaking depends on several factors besides the magnitude of the earthquake, one of the most important being soil conditions. For instance, thick layers of soft soil (such as fill) can amplify seismic waves, often at a considerable distance from the source, while sedimentary basins will often resonate, increasing the duration of shaking. This is why, in the 1989 Loma Prieta earthquake, the Marina district of San Francisco was one of the most damaged areas, though it was nearly 100 km from the epicenter. Geological structures were also significant, such as where seismic waves passing under the south end of San Francisco Bay reflected off the base of the Earth's crust towards San Francisco and Oakland. A similar effect channeled seismic waves between the other major faults in the area.

Magnitude Scales

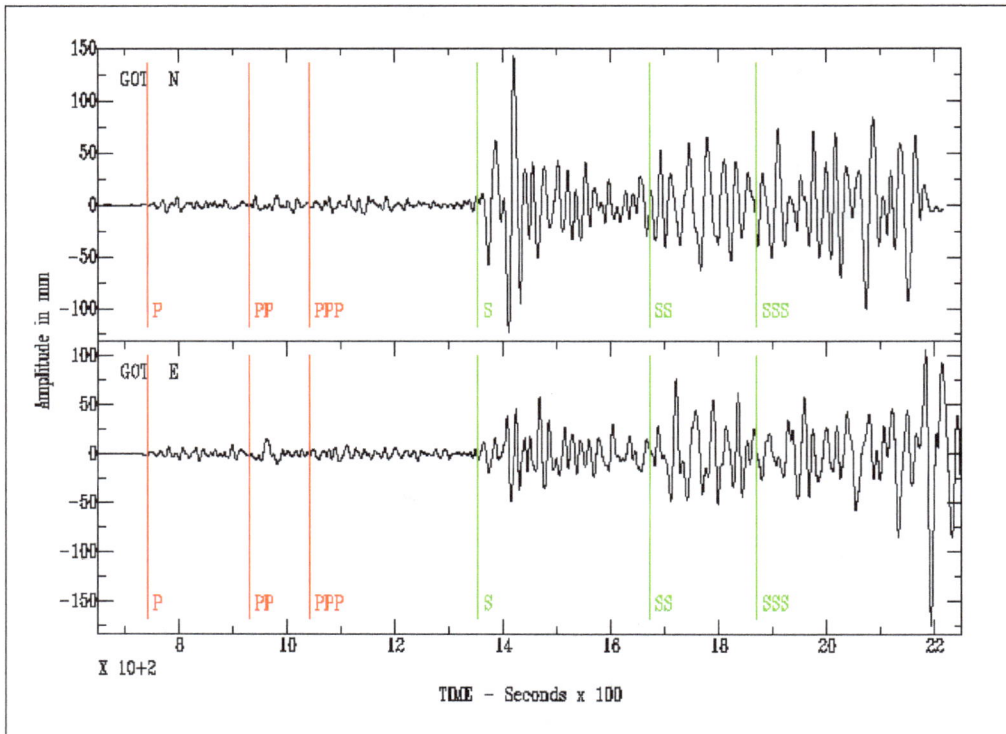

The compressive P-waves (following the red lines) – essentially sound passing through rock – are the fastest seismic waves, and arrive first, typically in about 10 seconds for an earthquake around 50 km away. The sideways-shaking S-waves (following the green lines) arrive some seconds later, traveling a little over half the speed of the P-waves; the delay is a direct indication of the distance to the quake. S-waves may take an hour to reach a point 1000 km away. Both of these are body-waves, that pass directly through the earth's crust. Following the S-waves are various kinds of surface-waves – Love waves and Rayleigh waves – that travel only at the earth's surface. Surface waves are smaller for deep earthquakes, which have less interaction with the surface. For shallow earthquakes – less than roughly 60 km deep – the surface waves are stronger, and may last several minutes; these carry most of the energy of the quake, and cause the most severe damage.

An earthquake radiates energy in the form of different kinds of seismic waves, whose characteristics reflect the nature of both the rupture and the earth's crust the waves travel through. Determination of an earthquake's magnitude generally involves identifying specific kinds of these waves on a seismogram, and then measuring one or more characteristics of a wave, such as its timing, orientation, amplitude, frequency, or duration. Additional adjustments are made for distance, kind of crust, and the characteristics of the seismograph that recorded the seismogram.

The various magnitude scales represent different ways of deriving magnitude from such information as is available. All magnitude scales retain the logarithmic scale as devised by Charles Richter, and are adjusted so the mid-range approximately correlates with the original "Richter" scale.

Most magnitude scales are based on measurements of only part of an earthquake's seismic wavetrain, and therefore are incomplete. This results in systematic underestimation of magnitude in certain cases, a condition called saturation.

Since 2005 the International Association of Seismology and Physics of the Earth's Interior (IAS-PEI) has standardized the measurement procedures and equations for the principal magnitude scales, M_L, M_s, mb, mB and mb_{Lg}.

Richter Magnitude Scale

The first scale for measuring earthquake magnitudes, developed in 1935 by Charles F. Richter and popularly known as the Richter scale, is actually the Local magnitude scale, label ML or M_L. Richter established two features now common to all magnitude scales. First, the scale is logarithmic, so that each unit represents a ten-fold increase in the *amplitude* of the seismic waves. As the energy of a wave is $10^{1.5}$ times its amplitude, each unit of magnitude represents a nearly 32-fold increase in the *energy* (strength) of an earthquake.

Second, Richter arbitrarily defined the zero point of the scale to be where an earthquake at a distance of 100 km makes a maximum horizontal displacement of 0.001 millimeters (1 μm, or 0.00004 in.) on a seismogram recorded with a Wood-Anderson torsion seismograph. Subsequent magnitude scales are calibrated to be approximately in accord with the original Richter (local) scale around magnitude 6.

All Local (ML) magnitudes are based on the maximum amplitude of the ground shaking, without distinguishing the different seismic waves. They underestimate the strength:

- of distant earthquakes (over ~600 km) because of attenuation of the S-waves.

- of deep earthquakes because the surface waves are smaller.

- of strong earthquakes (over M ~7) because they do not take into account the duration of shaking.

The original Richter scale, developed in the geological context of Southern California and Nevada, was later found to be inaccurate for earthquakes in the central and eastern parts of the continent (everywhere east of the Rocky Mountains) because of differences in the continental crust. All these problems prompted the development of other scales.

Most seismological authorities, such as the United States Geological Survey, report earthquake magnitudes above 4.0 as moment magnitude (below), which the press describes as Richter magnitude.

Other Local Magnitude Scales

Richter's original local scale has been adapted for other localities. These may be labelled "ML", or with a lowercase "l", either Ml, or M_l. Whether the values are comparable depends on whether the local conditions have been adequately determined and the formula suitably adjusted.

Japanese Meteorological Agency Magnitude Scale

In Japan, for shallow (depth < 60 km) earthquakes within 600 km, the Japanese Meteorological Agency calculates a magnitude labeled MJMA, M_{JMA}, or M_J. (These should not be confused with moment magnitudes JMA calculates, which are labeled M_w(JMA) or $M^{(JMA)}$, nor with the Shindo intensity scale). JMA magnitudes are based (as typical with local scales) on the maximum amplitude of the ground motion; they agree "rather well" with the seismic moment magnitude M_w in the range of 4.5 to 7.5, but underestimate larger magnitudes.

Body-wave Magnitude Scales

Body-waves consist of P-waves that are the first to arrive, or S-waves, or reflections of either. Body-waves travel through rock directly.

mB Scale

The original "body-wave magnitude" – mB or m_B (uppercase "B") – was developed by Gutenberg and Gutenberg & Richter to overcome the distance and magnitude limitations of the M_L scale inherent in the use of surface waves. mB is based on the P- and S-waves, measured over a longer period, and does not saturate until around M 8. However, it is not sensitive to events smaller than about M 5.5. Use of mB as originally defined has been largely abandoned, now replaced by the standardized mB_{BB} scale.

mb Scale

The mb or mb scale (lowercase "m" and "b") is similar to mB , but uses only P-waves measured in the first few seconds on a specific model of short-period seismograph. It was introduced in the 1960s with the establishment of the World-Wide Standardized Seismograph Network (WWSSN); the short period improves detection of smaller events, and better discriminates between tectonic earthquakes and underground nuclear explosions.

Measurement of mb has changed several times. As originally defined by Gutenberg m_b was based on the maximum amplitude of waves in the first 10 seconds or more. However, the length of the period influences the magnitude obtained. Early USGS/NEIC practice was to measure mb on the first second (just the first few P-waves), but since 1978 they measure the first twenty seconds. The modern practice is to measure short-period mb scale at less than three seconds, while the broadband mB_{BB} scale is measured at periods of up to 30 seconds.

mb_{Lg} Scale

The regional mb_{Lg} scale – also denoted mb_Lg, mbLg, MLg (USGS), Mn, and m_N – was developed by Nuttli for a problem the original M_L scale could not handle: All of North America east of the Rocky Mountains. The M_L scale was developed in southern California, which lies on blocks of oceanic crust, typically basalt or sedimentary rock, which have been accreted to the continent. East of the Rockies the continent is a craton, a thick and largely stable mass of continental crust that is largely granite, a harder rock with different seismic characteristics. In this area the M_L scale gives anomalous results for earthquakes which by other measures seemed equivalent to quakes in California.

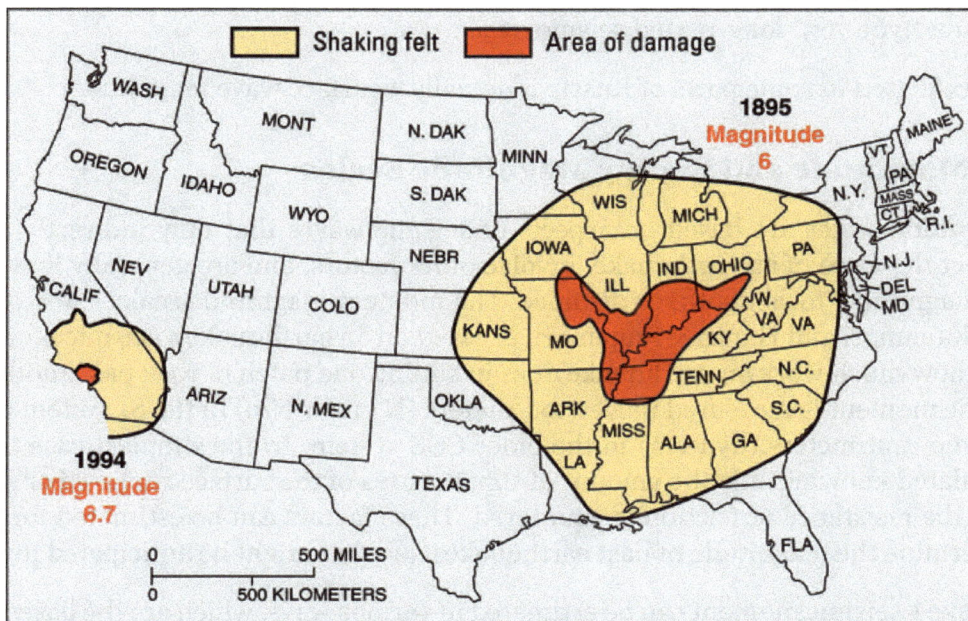

Differences in the crust underlying North America east of the Rocky Mountains makes that area more sensitive to earthquakes. Shown here: The 1895 New Madrid earthquake, M 6, was felt through most of the central U.S., while the 1994 Northridge quake, though almost ten times stronger at M 6.7, was felt only in southern California.

Nuttli resolved this by measuring the amplitude of short-period (1 sec.) Lg waves, a complex form of the Love wave which, although a surface wave, he found provided a result more closely related the mb scale than the M_s scale. Lg waves attenuate quickly along any oceanic path, but propagate well through the granitic continental crust, and Mb_{Lg} is often used in areas of stable continental crust; it is especially useful for detecting underground nuclear explosions.

Surface-wave Magnitude Scales

Surface waves propagate along the Earth's surface, and are principally either Rayleigh waves or Love waves. For shallow earthquakes the surface waves carry most of the energy of the earthquake, and are the most destructive. Deeper earthquakes, having less interaction with the surface, produce weaker surface waves.

The surface-wave magnitude scale, variously denoted as Ms, M_s, and M_s, is based on a procedure developed by Beno Gutenberg in 1942 for measuring shallow earthquakes stronger or more distant than Richter's original scale could handle. Notably, it measured the amplitude of surface waves (which generally produce the largest amplitudes) for a period of "about 20 seconds". The M_s scale approximately agrees with M_L at 6, then diverges by as much as half a magnitude. A revision by Nuttli, sometimes labeled M_{Sn}, measures only waves of the first second.

A modification – the "Moscow-Prague formula" – was proposed in 1962, and recommended by the IASPEI in 1967; this is the basis of the standardized M_{s20} scale (Ms_20, $M_s(20)$). A "broad-band" variant (Ms_BB, $M_s(BB)$) measures the largest velocity amplitude in the Rayleigh-wave train for periods up to 60 seconds. The M_{S7} scale used in China is a variant of M_s calibrated for use with the Chinese-made "type 763" long-period seismograph.

The MLH scale used in some parts of Russia is actually a surface wave magnitude.

Moment Magnitude and Energy Magnitude Scales

Other magnitude scales are based on aspects of seismic waves that only indirectly and incompletely reflect the force of an earthquake, involve other factors, and are generally limited in some respect of magnitude, focal depth, or distance. The moment magnitude scale – Mw or Mw – developed by Kanamori and Hanks & Kanamori, is based on an earthquake's seismic moment, M0, a measure of how much work an earthquake does in sliding one patch of rock past another patch of rock. Seismic moment is measured in Newton-meters (N · m or Nm) in the SI system of measurement, or dyne-centimeters (dyn-cm) in the older CGS system. In the simplest case the moment can be calculated knowing only the amount of slip, the area of the surface ruptured or slipped, and a factor for the resistance or friction encountered. These factors can be estimated for an existing fault to determine the magnitude of past earthquakes, or what might be anticipated for the future.

An earthquake's seismic moment can be estimated in various ways, which are the bases of the M_{wb}, M_{wr}, M_{wc}, M_{ww}, M_{wp}, M_i, and M_{wpd} scales, all subtypes of the generic M_w scale.

Seismic moment is considered the most objective measure of an earthquake's size in regard of total energy. However, it is based on a simple model of rupture, and on certain simplifying assumptions; it incorrectly assumes that the proportion of energy radiated as seismic waves is the same for all earthquakes.

Much of an earthquake's total energy as measured by Mw is dissipated as friction (resulting in heating of the crust). An earthquake's potential to cause strong ground shaking depends on the comparatively small fraction of energy radiated as seismic waves, and is better measured on the energy magnitude scale, Me. The proportion of total energy radiated as seismic waves varies greatly depending on focal mechanism and tectonic environment; Me and Mw for very similar earthquakes can differ by as much as 1.4 units.

Despite the usefulness of the M_e scale, it is not generally used due to difficulties in estimating the radiated seismic energy.

Two Earthquakes Differing Greatly in the Damage Done

In 1997 there were two large earthquakes off the coast of Chile. The magnitude of the first, in July, was estimated at M_w 6.9, but was barely felt, and only in three places. In October a M_w 7.1 quake in nearly the same location, but twice as deep and on a different kind of fault, was felt over a broad area, injured over 300 people, and destroyed or seriously damaged over 10,000 houses. As can be seen in the table below, this disparity of damage done is not reflected in either the moment magnitude (M_w) or the surface-wave magnitude (M_s). Only when the magnitude is measured on the basis of the body-wave (mb) or the seismic energy (M_e) is there a difference comparable to the difference in damage.

Date	ISC #	Lat.	Long.	Depth	Damage	M_s	M_w	mb	M_e	Type of fault
06 July 1997	1035633	−30.06	−71.87	23 km	Barely felt	6.5	6.9	5.8	6.1	Interplate-thrust
15 Oct. 1997	1047434	−30.93	−71.22	58 km	Extensive	6.8	7.1	6.8	7.5	Intraslab-normal
Difference:						0.3	0.2	1.0	1.4	

Energy Class (*K*-Class) Scale

K is a measure of earthquake magnitude in the *energy class* or *K-class* system, developed in 1955 by Soviet seismologists in the remote Garm (Tadjikistan) region of Central Asia; in revised form it is still used for local and regional quakes in many states formerly aligned with the Soviet Union (including Cuba). Based on seismic energy ($K = \log E_S$, in Joules), difficulty in implementing it using the technology of the time led to revisions in 1958 and 1960. Adaptation to local conditions has led to various regional K scales, such as K_F and K_S.

K values are logarithmic, similar to Richter-style magnitudes, but have a different scaling and zero point. K values in the range of 12 to 15 correspond approximately to M 4.5 to 6. M(K), $M_{(K)}$, or possibly M_K indicates a magnitude M calculated from an energy class K.

Tsunami Magnitude Scales

Earthquakes that generate tsunamis generally rupture relatively slowly, delivering more energy at longer periods (lower frequencies) than generally used for measuring magnitudes. Any skew in the spectral distribution can result in larger, or smaller, tsunamis than expected for a nominal magnitude. The tsunami magnitude scale, M_t, is based on a correlation by Katsuyuki Abe of earthquake seismic moment (M_0) with the amplitude of tsunami waves as measured by tidal gauges. Originally intended for estimating the magnitude of historic earthquakes where seismic data is lacking but tidal data exist, the correlation can be reversed to predict tidal height from earthquake magnitude. (Not to be confused with the height of a tidal wave, or *run-up*, which is an intensity effect controlled by local topography). Under low-noise conditions, tsunami waves as little as 5 cm can be predicted, corresponding to an earthquake of M ~6.5.

Another scale of particular importance for tsunami warnings is the mantle magnitude scale, M_m. This is based on Rayleigh waves that penetrate into the Earth's mantle, and can be determined quickly, and without complete knowledge of other parameters such as the earthquake's depth.

Duration and Coda Magnitude Scales

M_d designates various scales that estimate magnitude from the *duration* or length of some part of the seismic wave-train. This is especially useful for measuring local or regional earthquakes, both powerful earthquakes that might drive the seismometer off-scale (a problem with the analog instruments formerly used) and preventing measurement of the maximum wave amplitude, and weak earthquakes, whose maximum amplitude is not accurately measured. Even for distant earthquakes, measuring the duration of the shaking (as well as the amplitude) provides a better measure of the earthquake's total energy. Measurement of duration is incorporated in some modern scales, such as M_{wpd} and mB_c.

M_c scales usually measure the duration or amplitude of a part of the seismic wave, the *coda*. For short distances (less than ~100 km) these can provide a quick estimate of magnitude before the quake's exact location is known.

Macroseismic Magnitude Scales

Magnitude scales generally are based on instrumental measurement of some aspect of the seismic wave as recorded on a seismogram. Where such records do not exist, magnitudes can be estimated from reports of the macroseismic events such as described by intensity scales.

One approach for doing this relates the maximum intensity observed (presumably this is over the epicenter), denoted I_o (capital I, subscripted zero), to the magnitude. It has been recommended that magnitudes calculated on this basis be labeled $M_w(I_o)$, but are sometimes labeled with a more generic M_{ms}.

Another approach is to make an *isoseismal map* showing the area over which a given level of intensity was felt. The size of the "felt area" can also be related to the magnitude. While the recommended label for magnitudes derived in this way is $M_o(An)$, the more commonly seen label is M_{fa}. A variant, M_{La}, adapted to California and Hawaii, derives the Local magnitude (M_L) from the size of the area affected by a given intensity. M_I has been used for moment magnitudes estimated from *isoseismal intensities* calculated per Johnston.

Peak ground velocity (PGV) and *Peak ground acceleration* (PGA) are measures of the force that causes destructive ground shaking. In Japan, a network of strong-motion accelerometers provides PGA data that permits site-specific correlation with different magnitude earthquakes. This correlation can be inverted to estimate the ground shaking at that site due to an earthquake of a given magnitude at a given distance. From this a map showing areas of likely damage can be prepared within minutes of an actual earthquake.

Other Magnitude Scales

Many earthquake magnitude scales have been developed or proposed, with some never gaining broad acceptance and remaining only as obscure references in historical catalogs of earthquakes. Other scales have been used without a definite name, often referred to as "the method of Smith", with the authors often revising their method. On top of this, seismological networks vary on how they measure seismograms. When the details of how a magnitude has been determined are unknown, catalogs will specify the scale as unknown (variously Unk, Ukn, or UK). In such cases, the magnitude is considered generic and approximate.

An M_h (magnitude determined by hand) label has been used where the magnitude is too small or the data too poor (typically from analog equipment) to determine a Local magnitude, or multiple shocks or cultural noise complicates the records. The Southern California Seismic Network uses this "magnitude" where the data fail the quality criteria.

A special case is the *Seismicity of the Earth* catalog of Gutenberg & Richter. Hailed as a milestone as a comprehensive global catalog of earthquakes with uniformly calculated magnitudes, they never published the full details of how they determined those magnitudes. Consequently, while some catalogs identify these magnitudes as M_{GR}, others use UK (meaning "computational method unknown"). Subsequent study found many of the M_s values to be "considerably overestimated." Further study has found that most of the M_{GR} magnitudes "are basically M_s for large shocks shallower than 40 km, but are basically mB for large shocks at depths of 40–60 km." Gutenberg and Richter also used an italic, non-bold "*M* without subscript"– also used as a generic magnitude, and not to be confused with the bold, non-italic M used for *moment magnitude* – and a "unified magnitude" *m*. While these terms (with various adjustments) were used in scientific articles into the 1970s, they are now only of historical interest. An ordinary capital "M" without subscript is often used to refer to magnitude generically, where an exact value or the specific scale used is not important.

MODIFIED MERCALLI INTENSITY SCALE

The Modified Mercalli intensity scale (MM or MMI), descended from Giuseppe Mercalli's Mercalli intensity scale of 1902, is a seismic intensity scale used for measuring the intensity of shaking produced by an earthquake. It measures the effects of an earthquake at a given location, distinguished from the earthquake's inherent *force* or *strength* as measured by seismic magnitude scales (such as the "M_w" magnitude usually reported for an earthquake). While shaking is driven by the seismic energy released by an earthquake, earthquakes differ in how much of their energy is radiated as seismic waves. Deeper earthquakes also have less interaction with the surface, and their energy is spread out across a larger area. Shaking intensity is localized, generally diminishing with distance from the earthquake's epicenter, but can be amplified in sedimentary basins and certain kinds of unconsolidated soils.

Intensity scales empirically categorize the intensity of shaking based on the effects reported by untrained observers, and are adapted for the effects that might be observed in a particular region. In not requiring instrumental measurements, they are useful for estimating the magnitude and location of historical (pre-instrumental) earthquakes: The greatest intensities generally correspond to the epicentral area, and their degree and extent (possibly augmented with knowledge of local geological conditions) can be compared with other local earthquakes to estimate the magnitude.

Modified Mercalli Intensity Scale

The lower degrees of the Modified Mercalli Intensity scale generally deal with the manner in which the earthquake is felt by people. The higher numbers of the scale are based on observed structural damage.

This table gives Modified Mercalli scale intensities that are typically observed at locations near the epicenter of the earthquake.

I. Not felt	Not felt except by very few under especially favorable conditions.
II. Weak	Felt only by a few people at rest, especially on upper floors of buildings.
III. Weak	Felt quite noticeably by people indoors, especially on upper floors of buildings. Many people do not recognize it as an earthquake. Standing motor cars may rock slightly. Vibrations similar to the passing of a truck. Duration estimated.
IV. Light	Felt indoors by many, outdoors by few during the day. At night, some awakened. Dishes, windows, doors disturbed; walls make cracking sound. Sensation like heavy truck striking building. Standing motor cars rocked noticeably.
V. Moderate	Felt by nearly everyone; many awakened. Some dishes, windows broken. Unstable objects overturned. Pendulum clocks may stop.
VI. Strong	Felt by all, many frightened. Some heavy furniture moved; a few instances of fallen plaster. Damage slight.
VII. Very strong	Damage negligible in buildings of good design and construction; slight to moderate in well-built ordinary structures; considerable damage in poorly built or badly designed structures; some chimneys broken.
VIII. Severe	Damage slight in specially designed structures; considerable damage in ordinary substantial buildings with partial collapse. Damage great in poorly built structures. Fall of chimneys, factory stacks, columns, monuments, walls. Heavy furniture overturned.
IX. Violent	Damage considerable in specially designed structures; well-designed frame structures thrown out of plumb. Damage great in substantial buildings, with partial collapse. Buildings shifted off foundations. Liquefaction.
X. Extreme	Some well-built wooden structures destroyed; most masonry and frame structures destroyed with foundations. Rails bent.
XI. Extreme	Few, if any, (masonry) structures remain standing. Bridges destroyed. Broad fissures in ground. Underground pipe lines completely out of service. Earth slumps and land slips in soft ground. Rails bent greatly.
XII. Extreme	Damage total. Waves seen on ground surfaces. Lines of sight and level distorted. Objects thrown upward into the air.

Correlation with Magnitude

Magnitude	Magnitude/Intensity comparison
1.0–3.0	I
3.0–3.9	II–III
4.0–4.9	IV–V

5.0–5.9	VI–VII
6.0–6.9	VII–IX
7.0 and higher	VIII or higher

The correlation between magnitude and intensity is far from total, depending upon several factors including the depth of the hypocenter, terrain, distance from the epicenter. For example, a 4.5 magnitude quake in Salta, Argentina, in 2011, that was 164 km deep had a maximum intensity of I, while a 2.2 magnitude event in Barrow in Furness, England, in 1865, about 1 km deep had a maximum intensity of VIII.

Estimating Site Intensity and its use in Seismic Hazard Assessment

Dozens of so-called intensity prediction equations have been published to estimate the macroseismic intensity at a location given the magnitude, source-to-site distance and, perhaps, other parameters (e.g. local site conditions). These are similar to ground motion prediction equations for the estimation of instrumental strong-motion parameters such as peak ground acceleration. A summary of intensity prediction equations is available. Such equations can be used to estimate the seismic hazard in terms of macroseismic intensity, which has the advantage of being more closely related to seismic risk than instrumental strong-motion parameters.

Correlation with Physical Quantities

The Mercalli scale is not defined in terms of more rigorous, objectively quantifiable measurements such as shake amplitude, shake frequency, peak velocity, or peak acceleration. Human-perceived shaking and building damages are best correlated with peak acceleration for lower-intensity events, and with peak velocity for higher-intensity events.

Comparison to the Moment Magnitude Scale

The effects of any one earthquake can vary greatly from place to place, so there may be many Mercalli intensity values measured for the same earthquake. These values can be best displayed using a contoured map of equal intensity, known as an isoseismal map. However, each earthquake has only one magnitude.

SEISMOMETER

A seismometer is an instrument that responds to ground motions, such as caused by earthquakes, volcanic eruptions, and explosions. Seismometers are usually combined with a timing device and a recording device to form a seismograph. The output of such a device—formerly recorded on paper or film, now recorded and processed digitally—is a seismogram. Such data is used to locate and characterize earthquakes, and to study the Earth's internal structure.

The first seismometer was made in China during the 2nd Century. The first Western description of the device comes from the French physicist and priest Jean de Hautefeuille in 1703. The modern seismometer was developed in the 19th century.

In December 2018, a seismometer was deployed on the planet Mars by the *InSight* lander, the first time a seismometer was placed onto the surface of another planet.

Ancient Era

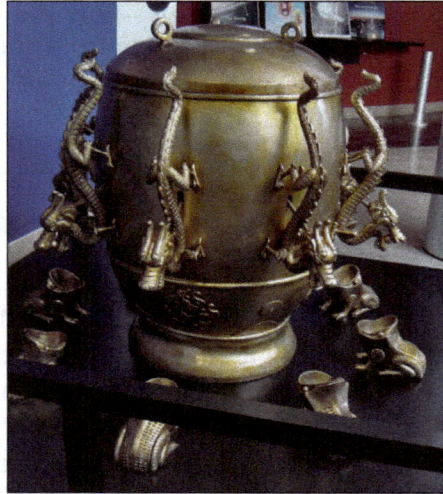

Replica of Zhang Heng's seismoscope *Houfeng Didong Yi.*

In AD 132, Zhang Heng of China's Han dynasty invented the first seismoscope, which was called *Houfeng Didong Yi* (translated as, "instrument for measuring the seasonal winds and the movements of the Earth"). The description we have, from the History of the Later Han Dynasty, says that it was a large bronze vessel, about 2 meters in diameter; at eight points around the top were dragon's heads holding bronze balls. When there was an earthquake, one of the dragons' mouths would open and drop its ball into a bronze toad at the base, making a sound and supposedly showing the direction of the earthquake. On at least one occasion, probably at the time of a large earthquake in Gansu in AD 143, the seismoscope indicated an earthquake even though one was not felt. The available text says that inside the vessel was a central column that could move along eight tracks; this is thought to refer to a pendulum, though it is not known exactly how this was linked to a mechanism that would open only one dragon's mouth. The first earthquake recorded by this seismoscope was supposedly "somewhere in the east". Days later, a rider from the east reported this earthquake.

Modern Designs

By the 13th century, seismographic devices existed in the Maragheh observatory in Persia. French physicist and priest Jean de Hautefeuille built one in 1703. After 1880, most seismometers were descended from those developed by the team of John Milne, James Alfred Ewing and Thomas Gray, who worked as foreign-government advisors in Japan from 1880 to 1895. These seismometers used damped horizontal pendulums. After World War II, these were adapted into the widely used Press-Ewing seismometer.

An early special-purpose seismometer consisted of a large, stationary pendulum, with a stylus on the bottom. As the earth started to move, the heavy mass of the pendulum had the inertia to stay still within the frame. The result is that the stylus scratched a pattern corresponding with the

Earth's movement. This type of strong-motion seismometer recorded upon a smoked glass (glass with carbon soot). While not sensitive enough to detect distant earthquakes, this instrument could indicate the direction of the pressure waves and thus help find the epicenter of a local quake. Such instruments were useful in the analysis of the 1906 San Francisco earthquake. Further analysis was performed in the 1980s, using these early recordings, enabling a more precise determination of the initial fault break location in Marin county and its subsequent progression, mostly to the south.

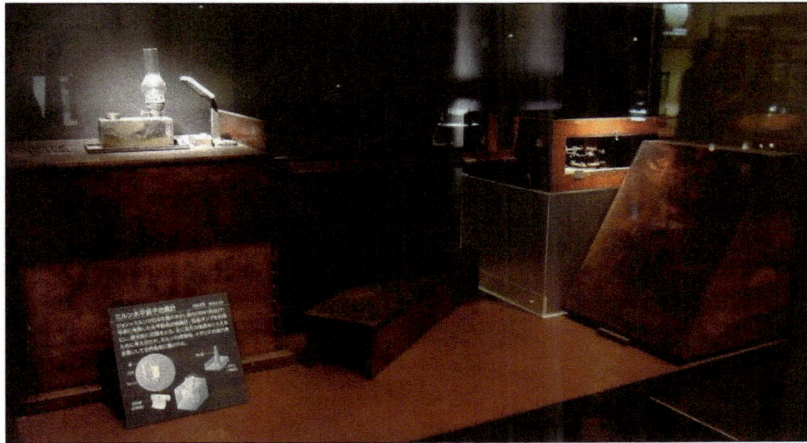

Milne horizontal pendulum seismometer.

Later, professional suites of instruments for the worldwide standard seismographic network had one set of instruments tuned to oscillate at fifteen seconds, and the other at ninety seconds, each set measuring in three directions. Amateurs or observatories with limited means tuned their smaller, less sensitive instruments to ten seconds. The basic damped horizontal pendulum seismometer swings like the gate of a fence. A heavy weight is mounted on the point of a long (from 10 cm to several meters) triangle, hinged at its vertical edge. As the ground moves, the weight stays unmoving, swinging the "gate" on the hinge.

The advantage of a horizontal pendulum is that it achieves very low frequencies of oscillation in a compact instrument. The "gate" is slightly tilted, so the weight tends to slowly return to a central position. The pendulum is adjusted (before the damping is installed) to oscillate once per three seconds, or once per thirty seconds. The general-purpose instruments of small stations or amateurs usually oscillate once per ten seconds. A pan of oil is placed under the arm, and a small sheet of metal mounted on the underside of the arm drags in the oil to damp oscillations. The level of oil, position on the arm, and angle and size of sheet is adjusted until the damping is "critical", that is, almost having oscillation. The hinge is very low friction, often torsion wires, so the only friction is the internal friction of the wire. Small seismographs with low proof masses are placed in a vacuum to reduce disturbances from air currents.

Zollner described torsionally suspended horizontal pendulums as early as 1869, but developed them for gravimetry rather than seismometry.

Early seismometers had an arrangement of levers on jeweled bearings, to scratch smoked glass or paper. Later, mirrors reflected a light beam to a direct-recording plate or roll of photographic paper. Briefly, some designs returned to mechanical movements to save money. In mid-twentieth-century systems, the light was reflected to a pair of differential electronic photosensors called a photomultiplier. The voltage generated in the photomultiplier was used to drive galvanometers which had a small mirror mounted on the axis. The moving reflected light beam would strike the surface of the turning drum, which was covered with photo-sensitive paper. The expense of developing photo sensitive paper caused many seismic observatories to switch to ink or thermal-sensitive paper.

Basic Principles

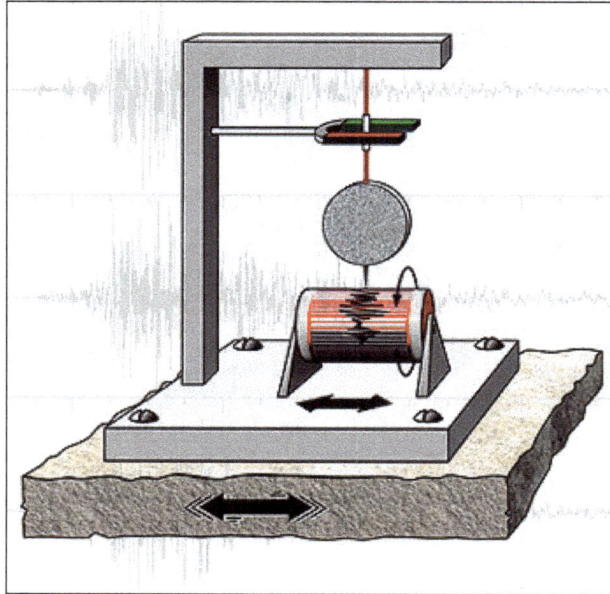

Basic horizontal-motion seismograph. The inertia of the round weight
tends to hold the pen still while the base moves back and forth.

A simple seismometer, sensitive to up-down motions of the Earth, is like a weight hanging from a spring, both suspended from a frame that moves along with any motion detected. The relative motion between the weight (called the mass) and the frame provides a measurement of the vertical ground motion. A rotating drum is attached to the frame and a pen is attached to the weight, thus recording any ground motion in a seismogram.

Any movement of the ground moves the frame. The mass tends not to move because of its inertia, and by measuring the movement between the frame and the mass, the motion of the ground can be determined.

Early seismometers used optical levers or mechanical linkages to amplify the small motions involved, recording on soot-covered paper or photographic paper. Modern instruments use electronics. In some systems, the mass is held nearly motionless relative to the frame by an electronic negative feedback loop. The motion of the mass relative to the frame is measured, and the feedback loop applies a magnetic or electrostatic force to keep the mass nearly motionless. The voltage needed to produce this force is the output of the seismometer, which is recorded digitally.

In other systems the weight is allowed to move, and its motion produces an electrical charge in a coil attached to the mass which voltage moves through the magnetic field of a magnet attached to the frame. This design is often used in a geophone, which is used in exploration for oil and gas.

Seismic observatories usually have instruments measuring three axes: North-South (y-axis), East-West (x-axis), and vertical (z-axis). If only one axis is measured, it is usually the vertical because it is less noisy and gives better records of some seismic waves.

The foundation of a seismic station is critical. A professional station is sometimes mounted on bedrock. The best mountings may be in deep boreholes, which avoid thermal effects, ground noise and tilting from weather and tides. Other instruments are often mounted in insulated enclosures on small buried piers of unreinforced concrete. Reinforcing rods and aggregates would distort the pier as the temperature changes. A site is always surveyed for ground noise with a temporary installation before pouring the pier and laying conduit. Originally, European seismographs were placed in a particular area after a destructive earthquake. Today, they are spread to provide appropriate coverage (in the case of weak-motion seismology) or concentrated in high-risk regions (strong-motion seismology).

Modern Instruments

CMG-40T triaxial broadband seismometer.

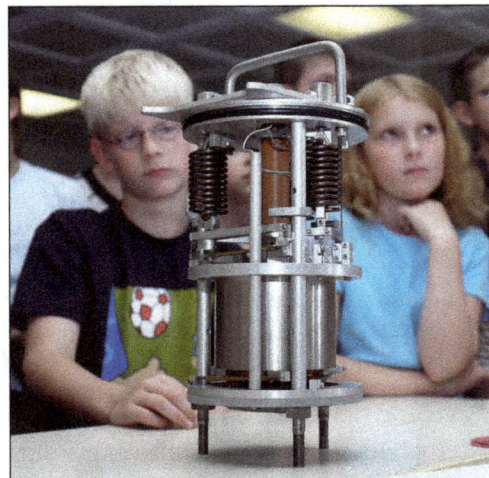

Seismometer without housing; presented during a demonstration
for children about earthquakes.

Modern instruments use electronic sensors, amplifiers, and recording devices. Most are broadband covering a wide range of frequencies. Some seismometers can measure motions with frequencies from 500 Hz to 0.00118 Hz (1/500 = 0.002 seconds per cycle, to 1/0.00118 = 850 seconds per cycle). The mechanical suspension for horizontal instruments remains the garden-gate described above. Vertical instruments use some kind of constant-force suspension, such as the LaCoste suspension. The LaCoste suspension uses a zero-length spring to provide a long period (high sensitivity). Some modern instruments use a "triaxial" design, in which three identical motion sensors are set at the same angle to the vertical but 120 degrees apart on the horizontal. Vertical and horizontal motions can be computed from the outputs of the three sensors.

Seismometers unavoidably introduce some distortion into the signals they measure, but professionally designed systems have carefully characterized frequency transforms.

Modern sensitivities come in three broad ranges: geophones, 50 to 750 V/m; local geologic seismographs, about 1,500 V/m; and teleseismographs, used for world survey, about 20,000 V/m. Instruments come in three main varieties: short period, long period and broadband. The short and long period measure velocity and are very sensitive, however they 'clip' the signal or go off-scale for ground motion that is strong enough to be felt by people. A 24-bit analog-to-digital conversion channel is commonplace. Practical devices are linear to roughly one part per million.

Delivered seismometers come with two styles of output: analog and digital. Analog seismographs require analog recording equipment, possibly including an analog-to-digital converter. The output of a digital seismograph can be simply input to a computer. It presents the data in a standard digital format (often "SE2" over Ethernet).

Teleseismometers

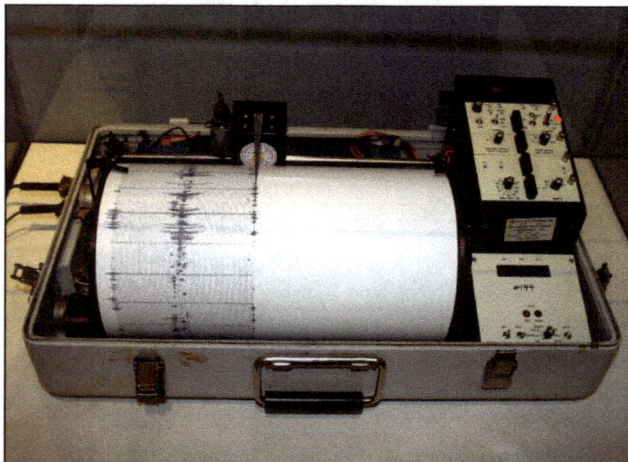

A low-frequency 3-direction ocean-bottom seismometer (cover removed). Two masses for x- and y-direction can be seen, the third one for z-direction is below. This model is a CMG-40TOBS, manufactured by Güralp Systems Ltd and is part of the Monterey Accelerated Research System.

The modern broadband seismograph can record a very broad range of frequencies. It consists of a small "proof mass", confined by electrical forces, driven by sophisticated electronics. As the earth moves, the electronics attempt to hold the mass steady through a feedback circuit. The amount of force necessary to achieve this is then recorded.

In most designs the electronics holds a mass motionless relative to the frame. This device is called a "force balance accelerometer". It measures acceleration instead of velocity of ground movement. Basically, the distance between the mass and some part of the frame is measured very precisely, by a linear variable differential transformer. Some instruments use a linear variable differential capacitor.

That measurement is then amplified by electronic amplifiers attached to parts of an electronic negative feedback loop. One of the amplified currents from the negative feedback loop drives a coil very like a loudspeaker, except that the coil is attached to the mass, and the magnet is mounted on the frame. The result is that the mass stays nearly motionless.

Most instruments measure directly the ground motion using the distance sensor. The voltage generated in a sense coil on the mass by the magnet directly measures the instantaneous velocity of the ground. The current to the drive coil provides a sensitive, accurate measurement of the force between the mass and frame, thus measuring directly the ground's acceleration (using f=ma where f=force, m=mass, a=acceleration).

One of the continuing problems with sensitive vertical seismographs is the buoyancy of their masses. The uneven changes in pressure caused by wind blowing on an open window can easily change the density of the air in a room enough to cause a vertical seismograph to show spurious signals. Therefore, most professional seismographs are sealed in rigid gas-tight enclosures. For example, this is why a common Streckeisen model has a thick glass base that must be glued to its pier without bubbles in the glue.

It might seem logical to make the heavy magnet serve as a mass, but that subjects the seismograph to errors when the Earth's magnetic field moves. This is also why seismograph's moving parts are constructed from a material that interacts minimally with magnetic fields. A seismograph is also sensitive to changes in temperature so many instruments are constructed from low expansion materials such as nonmagnetic invar.

The hinges on a seismograph are usually patented, and by the time the patent has expired, the design has been improved. The most successful public domain designs use thin foil hinges in a clamp.

Another issue is that the transfer function of a seismograph must be accurately characterized, so that its frequency response is known. This is often the crucial difference between professional and amateur instruments. Most instruments are characterized on a variable frequency shaking table.

Strong-motion Seismometers

Another type of seismometer is a digital strong-motion seismometer, or accelerograph. The data from such an instrument is essential to understand how an earthquake affects man-made structures, through earthquake engineering. The recordings of such instruments are crucial for the assessment of seismic hazard, through engineering seismology.

A strong-motion seismometer measures acceleration. This can be mathematically integrated later to give velocity and position. Strong-motion seismometers are not as sensitive to ground motions as teleseismic instruments but they stay on scale during the strongest seismic shaking.

Strong motion sensors are used for intensity meter applications.

Other Forms

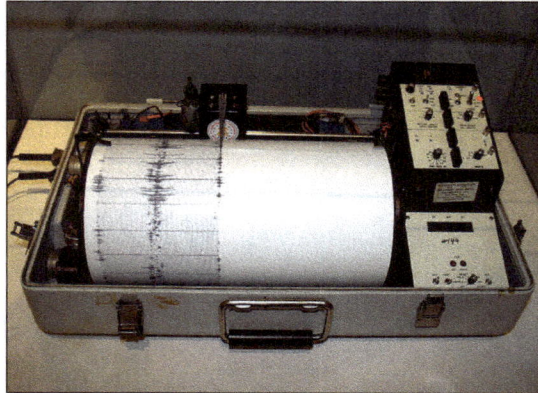

A Kinemetrics seismograph, formerly used by the United States Department of the Interior.

Accelerographs and geophones are often heavy cylindrical magnets with a spring-mounted coil inside. As the case moves, the coil tends to stay stationary, so the magnetic field cuts the wires, inducing current in the output wires. They receive frequencies from several hundred hertz down to 1 Hz. Some have electronic damping, a low-budget way to get some of the performance of the closed-loop wide-band geologic seismographs.

Strain-beam accelerometers constructed as integrated circuits are too insensitive for geologic seismographs, but are widely used in geophones.

Some other sensitive designs measure the current generated by the flow of a non-corrosive ionic fluid through an electret sponge or a conductive fluid through a magnetic field.

Interconnected Seismometers

Seismometers spaced in an array can also be used to precisely locate, in three dimensions, the source of an earthquake, using the time it takes for seismic waves to propagate away from the hypocenter, the initiating point of fault rupture. Interconnected seismometers are also used, as part of the International Monitoring System to detect underground nuclear test explosions, as well as for Earthquake early warning systems. These seismometers are often used as part of a large scale governmental or scientific project, but some organizations such as the Quake-Catcher Network, can use residential size detectors built into computers to detect earthquakes as well.

In reflection seismology, an array of seismometers image sub-surface features. The data are reduced to images using algorithms similar to tomography. The data reduction methods resemble those of computer-aided tomographic medical imaging X-ray machines (CAT-scans), or imaging sonars.

A worldwide array of seismometers can actually image the interior of the Earth in wave-speed and transmissivity. This type of system uses events such as earthquakes, impact events or nuclear explosions as wave sources. The first efforts at this method used manual data reduction from paper seismograph charts. Modern digital seismograph records are better adapted to direct computer use. With inexpensive seismometer designs and internet access, amateurs and small institutions have even formed a "public seismograph network".

Seismographic systems used for petroleum or other mineral exploration historically used an explosive and a wireline of geophones unrolled behind a truck. Now most short-range systems use "thumpers" that hit the ground, and some small commercial systems have such good digital signal processing that a few sledgehammer strikes provide enough signal for short-distance refractive surveys. Exotic cross or two-dimensional arrays of geophones are sometimes used to perform three-dimensional reflective imaging of subsurface features. Basic linear refractive geomapping software (once a black art) is available off-the-shelf, running on laptop computers, using strings as small as three geophones. Some systems now come in an 18" (0.5 m) plastic field case with a computer, display and printer in the cover.

Small seismic imaging systems are now sufficiently inexpensive to be used by civil engineers to survey foundation sites, locate bedrock, and find subsurface water.

Fiber Optic Cables as Seismometers

A new technique for detecting earthquakes has been found, using fiber optic cables. In 2016 a team of metrologists running frequency metrology experiments in England observed noise with a wave-form resembling the seismic waves generated by earthquakes. This was found to match seismological observations of an M_w 6.0 earthquake in Italy, ~1400 km away. Further experiments in England, Italy, and with a submarine fiber optic cable to Malta detected additional earthquakes, including one 4,100 km away, and an M_L 3.4 earthquake 89 km away from the cable.

Seismic waves are detectable because they cause micrometer-scale changes in the length of the cable. As the length changes so does the time it takes a packet of light to traverse to the far end of the cable and back (using a second fiber). Using ultra-stable metrology-grade lasers, these extremely minute shifts of timing (on the order of femtoseconds) appear as phase-changes.

The point of the cable first disturbed by an earthquake's p-wave (essentially a sound wave in rock) can be determined by sending packets in both directions in the looped pair of optical fibers; the difference in the arrival times of the first pair of perturbed packets indicates the distance along the cable. This point is also the point closest to the earthquake's epicenter, which should be on a plane perpendicular to the cable. The difference between the p-wave/s-wave arrival times provides a distance (under ideal conditions), constraining the epicenter to a circle. A second detection on a non-parallel cable is needed to resolve the ambiguity of the resulting solution. Additional observations constrain the location of the earthquake's epicenter, and may resolve the depth.

This technique is expected to be a boon in observing earthquakes, especially the smaller ones, in vast portions of the global ocean where there are no seismometers, and at a cost much cheaper than ocean bottom seismometers.

Recording

Today, the most common recorder is a computer with an analog-to-digital converter, a disk drive and an internet connection; for amateurs, a PC with a sound card and associated software is adequate. Most systems record continuously, but some record only when a signal is detected, as shown by a short-term increase in the variation of the signal, compared to its long-term average (which can vary slowly because of changes in seismic noise), also known as a STA/LTA trigger.

Viewing of a Develocorder film.

Matsushiro Seismological Observatory.

Prior to the availability of digital processing of seismic data in the late 1970s, the records were done in a few different forms on different types of media. A "Helicorder" drum was a device used to record data into photographic paper or in the form of paper and ink. A "Develocorder" was a machine that record data from up to 20 channels into a 16-mm film. The recorded film can be viewed by a machine. The reading and measuring from these types of media can be done by hand. After the digital processing has been used, the archives of the seismic data were recorded in magnetic tapes. Due to the deterioration of older magnetic tape medias, large number of waveforms from the archives are not recoverable.

Seismic Array

A seismic array is a system of linked seismometers arranged in a regular geometric pattern (cross, circle, rectangular etc.) to increase sensitivity to earthquake and explosion detection. A seismic array differs from a local network of seismic stations mainly by the techniques used for data analysis. The data from a seismic array is obtained using special digital signal processing techniques such as beamforming, which suppress noises and thus enhance the signal-to-noise ratio (SNR).

The earliest seismic arrays were built in the 1950s in order to improve the detection of nuclear tests worldwide. Many of these deployed arrays were classified until the 1990s. Today they become part of the IMS as primary or auxiliary stations. Seismic arrays are not only used to monitor earthquakes and nuclear tests, but also used as a tool for investigating nature and source regions of microseisms as well as locating and tracking volcanic tremor and analyzing complex seismic wave-field properties in volcanic areas.

Layout

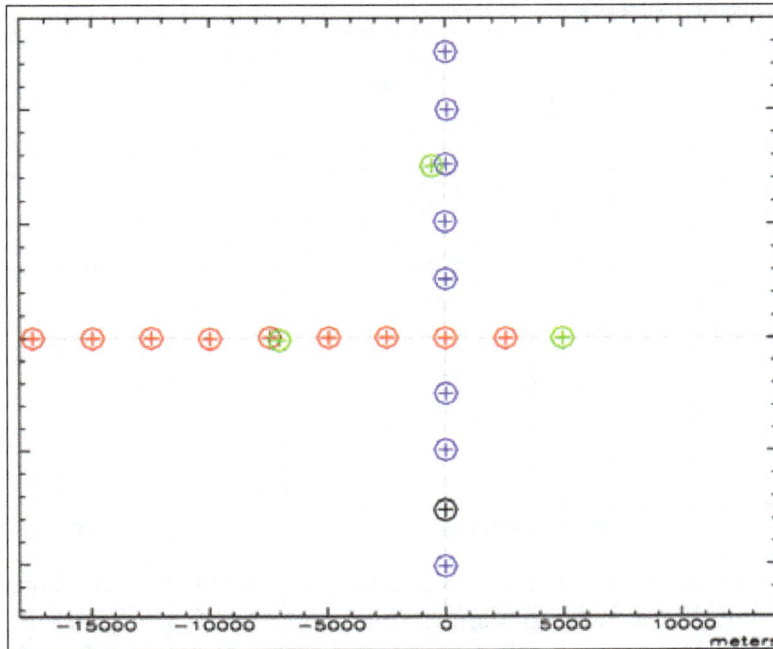

Layout of Yellowknife Seismological Array (YKA). Shortband seismometers are installed on blue and red sites, while broadband seismometers are installed on green sites.

Seismic arrays can be classified by size, which is defined by the array's aperture given by the largest distance between the single seismometers.

The sensors in a seismic array are arranged in different geometry patterns horizontally. The arrays built in the early 1960s were either cross (orthogonal linear) or L-shaped. The aperture of these arrays ranges from 10 to 25 km. Modern seismic arrays such as NORES and ARCES are located on concentric rings spaced at log-periodic intervals. Each ring consists of an odd number of seismometer sites. The number of rings and aperture differ from array to array, determined by economy and purpose.

Take the NORES design as an example, seismometers are placed on 4 concentric rings. The radii of the 4 rings are given by:

$$R_n = R_{min} \cdot 2.15^n \, (n = 0, 1, 2, 3), \, R_{min} = 150m$$

If the three sites in the inner ring are placed at 36, 156 and 276 degrees from due North, the five sites in the outer ring might be placed at 0, 72, 144, 216 and 288 degrees. This class of design is considered to provide the best overall array gain.

Data Processing

Array Beamforming

With a seismic array the signal-to-noise ratio (SNR) of a seismic signal can be improved by summing the coherent signals from the single array sites. The most important point during the beamforming process is to find the best delay times, with which the single traces must be shifted before summation in order to get the largest amplitudes due to coherent interference of the signals.

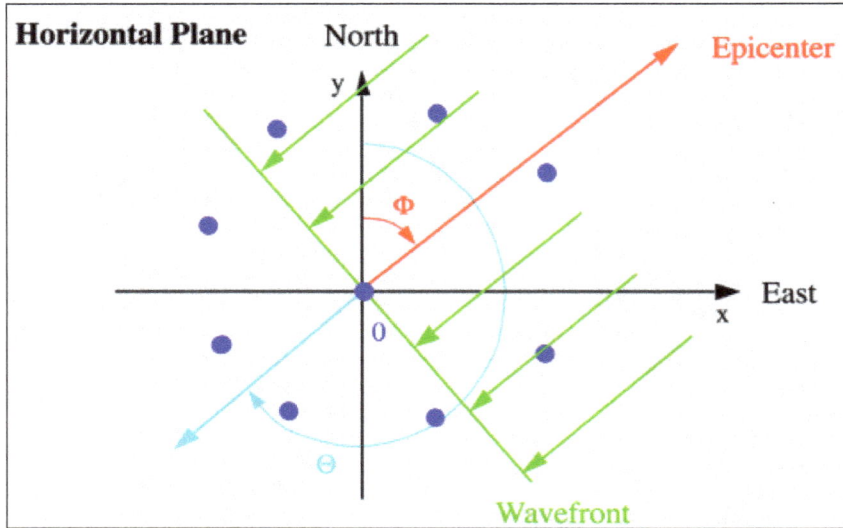

A wavefront coming from north-east and crossing a seismic array.

For distances from the source much larger than about 10 wavelengths, a seismic wave approaches an array as a wavefront that is close to planar. The directions of approach and propagation of the wavefront projected onto the horizontal plane are defined by the angles Φ and Θ.

- Φ Backazimuth (BAZ) = angle of wavefront approach, measured clockwise from the North to the direction towards the epicenter in degree.

- Θ Direction in which the wavefront propagates, measured in degree from the North, with $\Theta = \Phi \pm 180°$.

- d_j Horizontal distances between array site j and center site in [km].

- s Slowness vector with absolute value $s = 1/v_{app}$.

- v_{app} Apparent velocity vector with the absolute value $v_{app} = 1/s$. $v_{app} = (v_{app,x}, v_{app,y}, v_{app,z})$, where $v_{app,x}, v_{app,y}, v_{app,z}$ are the single apparent velocity components in [km/s] of the wavefront crossing an array.

- $v_{app,h}$ Absolute value of the horizontal component of the apparent velocity.

$$v_{app,h} = \sqrt{v_{app,x}^2 + v_{app,y}^2}$$

In most cases, the elevation differences between single array sites are so small that travel-time differences due to elevation differences are negligible. In this case, we cannot measure the vertical component of the wavefront propagation. The time delay τ_j between the center site o and site j with the relative coordinates (x_j, y_j) is:

$$\tau_j = \frac{d_j}{v_{app,h}} = \frac{-x_j sin\Phi - y_j cos\Phi}{v_{app,h}}$$

In some cases, not all array sites are located on one horizontal plane. The time delays τ_j also depends on the local crustal velocities (v_c) below the given site j. The calculation of τ_j with coordinates (x_j, y_j, z_j) is:

$$\tau_j = \frac{-x_j sin\Phi - y_j cos\Phi}{v_{app,h}} + \frac{z_j cos\Phi}{v_c}$$

In both the calculation can be written in vector syntax with position vector r_j and slowness vector s_j:

$$\tau_j = r_j \cdot s_j$$

Let $w_j(t)$ be the digital sample of the seismometer from site j at time t, then the beam of the whole array is defined as:

$$b(t) = \frac{1}{M} \sum_{j=1}^{M} w_j(t + \tau_j)$$

If seismic waves are harmonic waves S(t) without noise, with identical site responses, and without attenuation, then the above operation would reproduce the signal S(t) accurately. Real data w(t) are the sum of background noise n(t) plus the signal of interest S(t), i.e. w(t) = S(t) + n(t). Assuming that the signal is coherent and not attenuated, calculating the sum of M observations and including noise we get:

$$B(t) = M \cdot S(t) + \sum_{j=1}^{M} n_j(t + \tau_j)$$

Assuming that the noise $n_j(t)$ has a normal amplitude distribution with zero mean and variance σ^2 at all sites, then the variance of the noise after summation is $\sigma_s^2 = M\sigma^2$ and the standard deviation is $\sqrt{M\sigma^2}$. That means the standard deviation of the noise is multiplied by $\sqrt{M}$ while the coherent signal is multiplied by M. The theoretical improvement of the SNR by beamforming (aka array gain) will be $G = \sqrt{M}$ for an array containing M sites.

The N-th Root Process

N-th root process is a non-linear method to enhance the SNR during beamforming. Before summing up the single seismic traces, the N-th root is calculated for each trace retaining the sign information. signum{$w_j(t)$} is a function defined as -1 or +1, depending on the sign of the actual sample $w_j(t)$. N is an integer that has to be chosen by the analyst:

$$B_N(t) = \sum_{j=1}^{M} \sqrt[N]{n_j(t + \tau_j)} \cdot signum\{w_j(t)\}$$

Here the value of the function $\{w_j(t)\}$ is defined as ±1 depending on the sign of the actual sample w_j(t). After this summation, the beam has to be raised to the power of N:

$$b(t) = |B_N(t)|^N \cdot signum\{w_j(t)\}$$

N-th root process was first proposed by K. J. Muirhead and Ram Dattin in 1976. With N-th root process, the suppression of uncorrelated noise is better than with linear beamforming. However, it weights the coherency of a signal higher than the amplitudes, which results in a distortion of the waveforms.

Weighted Stack Methods

Schimmel and Paulssen introduced another non-linear stacking technique in 1997 to enhance signals through the reduction of incoherent noise, which shows a smaller waveform distortion than the N-th root process. Kennett proposed the use of the semblance of the signal as a weighting function in 2000 and achieved a similar resolution.

An easily implementable weighted stack method would be to weight the amplitudes of the single sites of an array with the SNR of the signal at this site before beamforming, but this does not directly exploit the coherency of the signals across the array. All weighted stack methods can increase the slowness resolution of velocity spectrum analysis.

Double Beam Technique

A cluster of earthquakes can be used as a source array to analyze coherent signals in the seismic coda. This idea was consequently expanded by Krüger in 1993 by analyzing seismic array data from well-known source locations with the so-called "double beam method". The principle of reciprocity is used for source and receiver arrays to further increase the resolution and the SNR for small amplitude signals by combining both arrays in a single analysis.

Array Transfer Function

The array transfer function describes sensitivity and resolution of an array for seismic signals with different frequency contents and slownesses. With an array, we are able to observe the wavenumber $k = 2\pi / \lambda = 2\pi \cdot f \cdot s$ of this wave defined by its frequency f and its slowness s. While time-domain analog-to-digital conversion may give aliasing effects in the time domain, the spatial sampling may give aliasing effects in the wavenumber domain. Thus the wavelength range of seismic signals and the sensitivity at different wavelengths must be estimated.

The difference between a signal w at the reference site A and the signal w_n at any other sensor A_n is the travel time between the arrivals at the sensors. A plane wave is defined by its slowness vector s_0:

$w_n(t) = w(t - r_n \cdot s_0)$, where r_n is the position vector of site n.

The best beam of an array with M sensors for a seismic signal for the slowness s_0 is defined as:

$$b(t) = \frac{1}{M} \sum_{j=1}^{M} w_j(t + r_j \cdot s_0)$$

If we calculate all time shifts for a signal with the slowness s_0 with respect to any other slowness s, the calculated beam becomes:

$$b(t) = \frac{1}{M} \sum_{j=1}^{M} w_j(t + r_j \cdot (s_0 - s))$$

The seismic energy of this beam can be calculated by integrating over the squared amplitudes,

$$E(t) = \int_{-\infty}^{\infty} b^2(t)dt = \int_{-\infty}^{\infty} [\frac{1}{M} \sum_{j=1}^{M} w_j(t + r_j \cdot (s_0 - s))]^2 dt$$

This equation can be written in the frequency domain with $\overline{w}(\omega)$ being the Fourier transform of the seismogram w(t), using the definition of the wavenumber vector k = $\omega \cdot$ s,

$$E(\omega, k_0 - k) = \frac{1}{2\pi} \int_{-\infty}^{\infty} |\overline{w}(\omega)|^2 \cdot |C(k_0 - k)|^2 d\omega ,$$

where,

$$C(k_0 - k) = \frac{1}{M} \sum_{j=1}^{M} e^{iwr_j(k_0-k)}$$

This equation is called the transfer function of an array. If the slowness difference is zero, the factor $|C(k_0 - k)|^2$ becomes 1.0 and the array is optimally tuned for this slowness. All other energy propagating with a different slowness will be suppressed.

Slowness Estimation

Slowness estimation is a matter of forming beams with different slowness vectors and comparing the amplitudes or the power of the beams, and finding out the best beam by looking for the v_{app} and backazimuth combination with the highest energy on the beam.

F-k Analysis

Frequency-wavenumber analysis is used as a reference tool in array processing for estimating slowness. This method was proposed by Capon in 1969 and further developed to include wide-band analysis, maximum-likelihood estimation techniques, and three-component data in the 1980s.

The methodology exploits the deterministic, non-periodic character of seismic wave propagation to calculate the frequency-wavenumber spectrum of the signals by applying the multidimensional Fourier transform. A monochromatic plane wave w(x,t) will propagate along the x direction according to equation:

$$w(x,t) = Ae^{i2\pi(f_0 t - k_0 x)}$$

It can be rewritten in frequency domain as:

$$W(k_x, f) = A\delta(f - f_0)\delta(k_x - k_0)$$

which suggests the possibility to map a monochromatic plane wave in the frequency-wavenumber domain to a point with coordinates $(f, k_x) = (f_0, k_0)$.

Practically, f-k analysis is performed in the frequency domain and represents in principle beam-forming in the frequency domain for a number of different slowness values. At NORSAR slowness values between -0.4 and 0.4 s/km are used equally spaced over 51 by 51 points. For every one of these points the beam power is evaluated, giving an equally spaced grid of 2601 points with power information.

Beampacking

A beampacking scheme was developed at NORSAR to apply f-k analysis of regional phases to data of large array. This algorithm performs time-domain beamforming over a predefined grid of slowness points and measures the power of the beam.

In practice the beampacking process gives the same slowness estimate as for the f-k analysis in the frequency domain. Compared to the f-k process, the beampacking process results in a slightly (about 10%) narrower peak for the maximum power.

Plane Wave Fitting

Another way of estimating slowness is to pick carefully times of the first onset or any other common distinguishable part of the same phase (same cycle) for all instruments in an array. Let t_i be the arrival time picked at site i, and t_{ref} be the arrival time at the reference site, then $\tau_i = t_i - t_{ref}$ is the observed time delay at site i. We observe the plane wave at M sites. With $M \geq 3$. The horizontal components (s_x, s_y) of the slowness vector s can be estimated by:

$$\hat{s} = \min_{s} \sum_{j=1}^{M}(\tau_j - r_j \cdot s)^2$$

Plane wave fitting requires interactive analyst's work. However, to obtain automatic time picks and thereby provide a slowness estimate automatically, techniques like cross-correlation or just picking of peak amplitude within a time window may be used. Because of the amount of required computations, plane wave fitting is most effective for arrays with a smaller number of sites or for subarray configurations.

Applications

YKA

YKA or Yellowknife Seismological Array is a medium size seismic array established near Yellowknife in the Northwest Territories, Canada, in 1962, in cooperative agreement between the Department of Mines and Technical Surveys (now Natural Resources Canada) and the United Kingdom Atomic Energy Authority (UKAEA), to investigate the feasibility of teleseismic detection and identification of nuclear explosions. YKA currently consists of 19 short period seismic sensors in the form of a cross with an aperture of 2.5 km, plus 4 broadband seismograph sites with instruments able to detect a wide range of seismic wave frequencies.

LASA

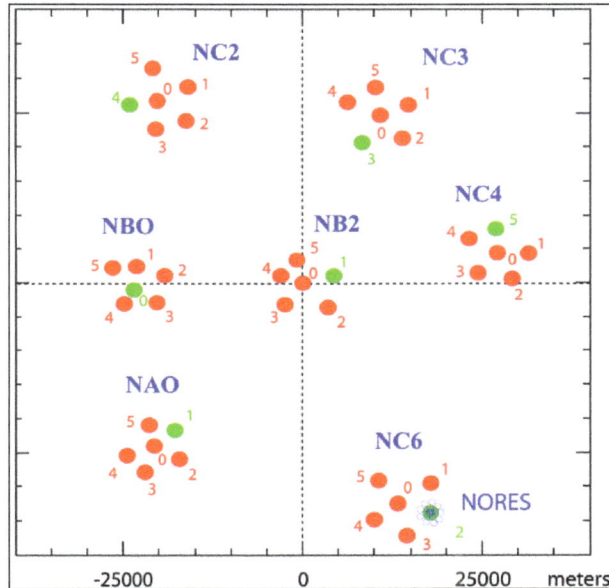

Configuration of large aperture array NORSAR and small aperture array NORES.

LASA or Large Aperture Seismic Array is the first large seismic array. It was built in Montana, USA, in 1965.

NORSAR

NORSAR or Norwegian Seismic Array was established at Kjeller, Norway in 1968 as part of the Norwegian-US agreement for the detection of earthquakes and nuclear explosions. It has been an independent, not-for-profit, research foundation within the field of geo-science since 1999. NOR-SAR was constructed as a large aperture array with a diameter of 100 km. It is the largest stand-alone array in the world.

NORES and ARCES

NORES was the first regional seismic array constructed in southern Norway in 1984. A sister array ARCES was established in northern Norway in 1987. NORES and ARCES are small aperture arrays with a diameter of only 3 km.

GERES

GERES is a small aperture array built in the Bavarian Forest near the border triangle of Germany, Austria and Czech, in 1988. It consists of 25 individual seismic stations arranged in 4 concentric rings with radius of 200m, 430m, 925m and 1988m.

SPITS

SPITS is a very small aperture array at Spitsbergen, Norway. It was originally installed in 1992 and upgraded to IMS standard in 2007 by NORSAR.

Ocean-bottom Seismometer

An ocean-bottom seismometer (OBS) is a seismometer that is designed to record the
earth motion under oceans and lakes from man-made sources and natural sources.

Sensors at the sea floor are used to observe acoustic and seismic events. Seismic and acoustic signals may be caused by different sources, by earthquakes and tremors as well as by artificial sources. Computing and analyzing the data yields information about the kind of source and, in case of natural seismic events, the geophysics and geology of the sea floor and the deeper crust. The deployment of OBS along a profile will give information about the deep structure of the Earth's crust and upper mantle in offshore areas. OBS may be equipped with a maximum of a three-component geophone in addition to a hydrophone, and thus it needs a capacity of more than 144 Mbytes, which would be the minimum for an adequate MCS profiling. In a typical survey, the instruments should be operational for several days (deployments can exceed 12 months), which requires a data storage capacity of more than 500 Mbyte. Other experiments, such as tomographic investigations within a 3D-survey or seismological monitoring, demand even larger capacities.

Instrument Package

The OBS consists of an aluminum sphere which contains sensors, electronics, enough alkaline batteries to last 10 days on the ocean bottom, and an acoustic release. The two sphere halves are put together with an O-ring and a metal clamp to hold the halves together. A slight vacuum is placed on the sphere to better ensure a seal. The sphere by itself floats, so an anchor is needed to sink the instrument to the bottom. In this case, the anchor is flat metal plate 40 inches (1.02 meters) in diameter. The instrument has been designed to be able to deploy and recover off almost any vessel. All that is needed (for deployment and recovery) is enough deck space to hold the instruments and their anchors and a boom capable of lifting an OBS off the deck and swing it over to lower it into the water. The OBS is bolted to the anchor and then dropped (gently) over the side.

An ocean-bottom seismometer goes over the side of R/V Oceanus. It will record long-period, low-frequency seismicwaves for up to a year before the ship returns to retrieve the instrument.

Working

Seismometers work using the principle of inertia. The seismometer body rests securely on the sea floor. Inside, a heavy mass hangs on a spring between two magnets. When the earth moves, so do the seismometer and its magnets, but the mass briefly stays where it is. As the mass oscillates through the magnetic field it produces an electric current which the instrument measures. The seismometer itself is a small metal cylinder; the rest of the footlocker-sized OBS consists of equipment to run the seismometer (a data logger and batteries), weight to sink it to the sea floor, a remote-controlled acoustic release and flotation to bring the instrument back to the surface.

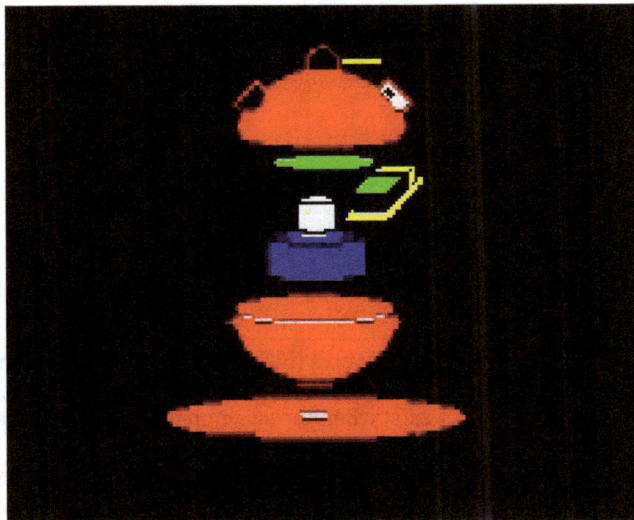

Types of OBS

The ground motion caused by earthquakes can be extremely small (less than a millimeter) or large (several meters). Small motions have high frequencies, so monitoring them requires measuring movement many times per second and produces huge amounts of data. Large motions are much rarer, so instruments need to record data less frequently, to save memory space and battery power for longer deployments. Because of this variability, engineers have designed two basic kinds of seismometers:

Short-period OBSs

They record high-frequency motions (up to hundreds of times per second). They can record small, short-period earthquakes and are also useful for studying the outer tens of kilometers of the seafloor. Technical details for two models: WHOI D2 and Scripps L-CHEAPO.

Long-period OBSs

They record a much broader range of motions, with frequencies of about 10 per second to once or twice a minute. They are used for recording mid-sized earthquakes and seismic activity far from the instrument. Technical details for two models: WHOI long-deployment OBS and Scripps long-deployment OBS.

Custom OBSs

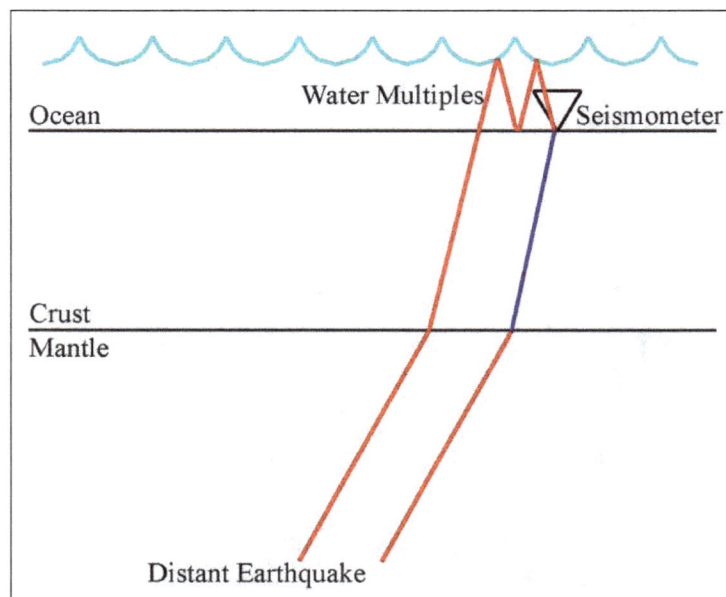

This shows a P wave (red) converting into an S wave (blue), with the P wave having the ability to travel through the ocean and reflect back down off of the surface into the seismometer. This creates water multiples that do not exist when the seismometer has free-air above it.

Custom OBS are beginning to be developed, as the need for expansion on coverage in the field of seismology increases and permanent deployments are necessary. One customizations to improve the data quality of the seismometers is to borehole the seismometer in an aluminum casing into the surface (~1 m) to create stability in the soft sediment of the ocean floor. Another customization that is possible is to add a differential pressure gauge (DPG) and/or current meter, in order to understand how the pressure is changing around the seismometer. It can also be practical to store the datalogger and battery in a glass Benthos sphere in order to be able to connect to the ship through the use of a remotely operated vehicle (ROV), which is a necessary advancement in order to have and maintain permanent OBS deployments.

Advantages

Very stable clocks make comparable the readings from many far-flung seismometers. (Without reliable time-stamps, data from different machines would be unusable.) Development of these clocks was a crucial advance for seismologists studying the Earth's interior. After recovering an ocean-bottom seismometer, scientists can offload the instrument's data by plugging in a data cable. This feature saves the task of gingerly disassembling the instrument's protective casing while aboard a rolling ship. The ability to connect a seismometer to a mooring or observatory makes the instrument's data instantly available. This is a huge advantage for geologists scrambling to respond to a major earthquake.

Disadvantages

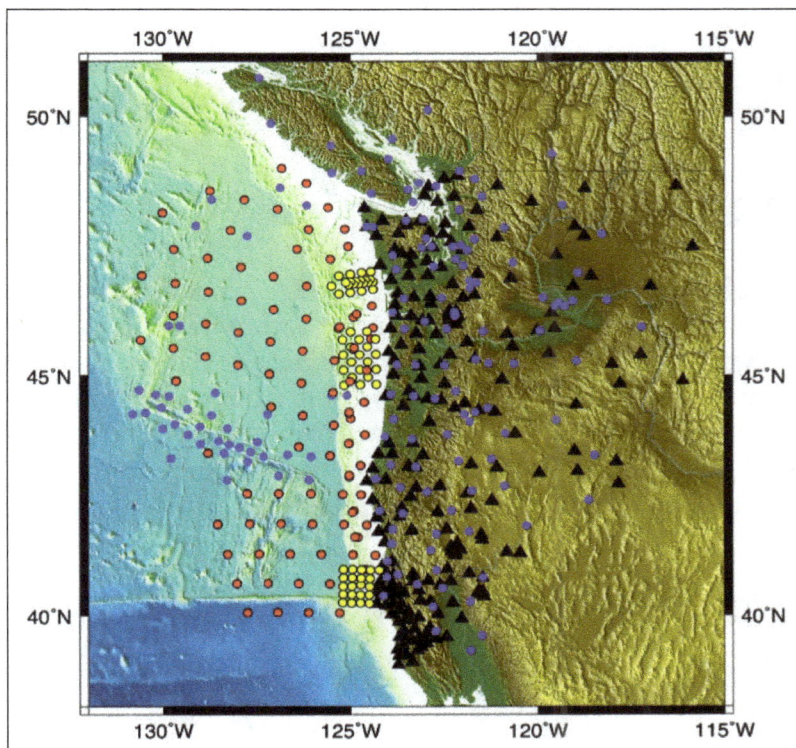

This is a map of the land and ocean-bottom stations that were deployed in the Cascadia Initiative.

The environment of these deployments complicates standard methods that are used in analyzing the data because of the ocean on top of the seismometer, as opposed to free-air above a typical land station. These seismometers also have a decreased signal-to-noise ratio because of noise created by the movement of the oceans due to wind driven tides, particularly at periods of 7 and 14 seconds. This long period motion and current flowing around the seismometer can create problems of long period noise on the horizontal components because the soft (saturated) sediment that the seismometer is resting on is more susceptible to allow the seismometer to tilt and ideally, the horizontal component will not move and be perpendicular to gravity to get the best results out of the seismometer. The saturated sediment also reduces signal-to-noise ratio significantly because the velocity of the P- and S-waves decreases and the seismic waves get trapped in the sediment layer creating a large amplitude ringing due to the conservation of energy.

Notable Deployments

One of the largest OBS deployments ever was The Big Mantle Electromagnetic and Tomography (Big MELT) Experiment, involving almost 100 OBS in the East Pacific Rise with the goal of understanding magma generation and mid-ocean ridge development. The Cascadia Initiative is an offshore/onshore deployment to observe the deformation of the Juan de Fuca and Gorda Plates, as well as topics ranging from megathrust earthquakes to volcanic arc structure in the Pacific Northwest. The Hawaiian PLUME (Plume-Lithosphere Undersea Melt Experiment) was an onshore/offshore (predominantly offshore) deployment to better understand what type of mantle plume is beneath Hawaii and to better understand the mantle upwelling in this region and its relationship to the lithosphere. The Asthenospheric and Lithospheric Broadband Architecture from the California Offshore Region Experiment (ALBACORE) deployment from 2010 to 2011 of 34 OBS to help better understand the tectonic interaction at the Pacific-North America plate boundary and deformation styles of the Pacific plate and the nearby microplates.

SEISMOGRAM

A seismogram is a graph output by a seismograph. It is a record of the ground motion at a measuring station as a function of time. Seismograms typically record motions in three cartesian axes (x, y, and z), with the z axis perpendicular to the Earth's surface and the x- and y- axes parallel to the surface. The energy measured in a seismogram may result from an earthquake or from some other source, such as an explosion. Seismograms can record lots of things, and record many little waves, called microseisms. These tiny microseisms can be caused by heavy traffic near the seismograph, waves hitting a beach, the wind, and any number of other ordinary things that cause some shaking of the seismograph.

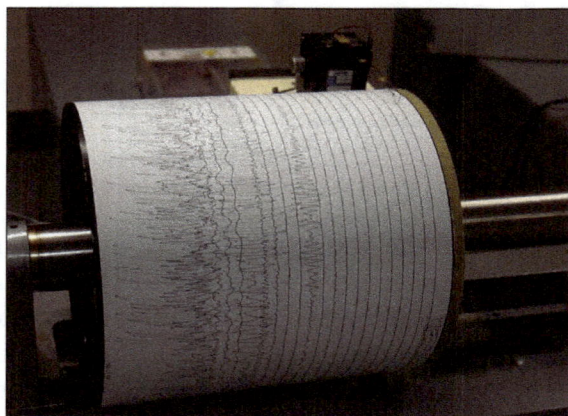

A seismogram being recorded by a seismograph at Weston Observatory in Massachusetts.

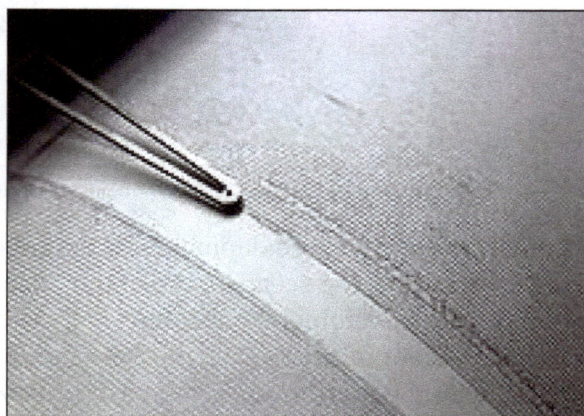

A detail of the seismogram.

Historically, seismograms were recorded on paper attached to rotating drums. Some used pens on ordinary paper, while others used light beams to expose photosensitive paper. Today, practically all seismograms are recorded digitally to make analysis by computer easier. Some drum seismometers are still found, especially when used for public display. Seismograms are essential for finding the location and magnitude of earthquakes.

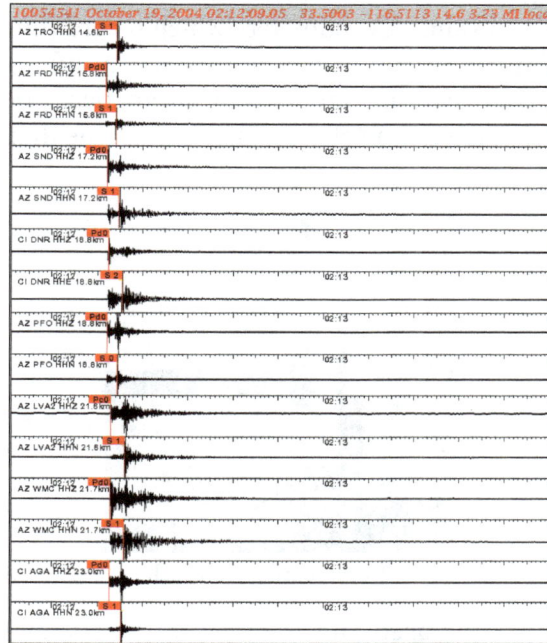

A set of seismograms for an earthquake.

Recording

Prior to the availability of digital processing of seismic data in the late 1970s, the records were done in a few different forms on different types of media.

A Helicorder drum is a device used to record data into photographic paper or in the form of paper and ink. A piece of paper is wrapped around a rotating drum of the helicorder which receives the seismic signal from a seismometer. For each predefined interval of data, the helicorder will plot the seismic data in one line before moving to the next line at the next interval. The paper must be changed after the helicorder writes on the last line of the paper. In the model that use ink, regular maintenance of the pen must be done for accurate recording.

A container that stores a Develocorder film reel.

A Develocorder is a machine that record multi-channels seismic data into a 16 mm film. The machine was developed by Teledyne Geotech during the mid 1960s. It can automatically plot seismograms from 18 seismic signal sources and 3 time signals on a continuous reel of film. The signals from seismometers are processed by 15.5 Hz recording galvanometers which record the seismograms to a reel of 200 feet (61 m) of film at the speeds between 3 and 20 centimetres (1.2 and 7.9 in) per minute. The machine has self-contained circulating chemicals that are used to automatically develop the film. However, the machine takes at least 10 minutes from time of recording to the time that the film can be viewed.

A Develocorder film reel.

After the digital processing has been used, the archives of the seismograms were recorded in magnetic tapes. The data from the magnetic tapes can then be read back to reconstruct the original waveforms. Unfortunately, due to the deterioration of older magnetic tape medias, large number of waveforms from the archives in the early digital recording days are not recoverable. Today, many other forms are used to digitally record the seismograms into digital medias.

Viewing of a Develocorder film.

Reading

Seismograms are read from left to right. Time marks show when the earthquake occurred. Time is shown by half-hour (thirty-minute) units. Each rotation of the seismograph drum is thirty minutes. Therefore, on seismograms, each line measures thirty minutes. This is a more efficient way to read a seismogram. Secondly, there are the minute-marks. A minute mark looks like a hyphen "-" between each minute. Minute marks count minutes on seismograms. From left to right, each mark stands for a minute.

Each seismic wave looks different. The P-wave is the first wave that is bigger than the other waves (the microseisms). Because P waves are the fastest seismic waves, they will usually be the first ones that the seismograph records. The next set of seismic waves on the seismogram will be the S-waves. These are usually bigger than the P waves, and have higher frequency. Look for a dramatic change in frequency for a different type of wave.

SEISMIC INTERFEROMETRY

Interferometry examines the general interference phenomena between pairs of signals in order to gain useful information about the subsurface. Seismic interferometry (SI) utilizes the crosscorrelation of signal pairs to reconstruct the impulse response of a given media. Papers by Keiiti Aki, Géza Kunetz, and Jon Claerbout helped develop the technique for *seismic* applications and provided the framework upon which modern theory is based.

A signal at a location A can be crosscorrelated with a signal at a location B to reproduce a virtual source-receiver pair using seismic interferometry. Crosscorrelation is often considered the key mathematical operation in this approach, but it is also possible to use convolution to come up with a similar result. The crosscorrelation of passive noise measured at a free surface reproduces the subsurface impulse response. As such, it is possible to obtain information about the subsurface with no need for an active seismic source. This method, however, is not limited to passive sources, and can be extended for use with active sources and computer–generated waveforms.

As of 2006 the field of seismic interferometry was beginning to change the way geophysicists view noise. Seismic interferometry uses this previously–ignored noise in models of the shallow subsurface. Potential applications include both research and industry.

Claerbout developed a workflow to apply existing interferometry techniques to investigating the shallow subsurface, although it was not proven until later that seismic interferometry could be applied to real world media. The long term average of random ultrasound waves can reconstruct the impulse response between two points on an aluminum block. However, they had assumed random diffuse noise, limiting interferometry in real world conditions. In a similar case, it was shown that the expressions for uncorrelated noise sources reduce to a single crosscorrelation of observations at two receivers. The interferometric impulse response of the subsurface can be reconstructed using only an extended record of background noise, initially only for the surface and direct wave arrivals.

Crosscorrelations of seismic signals from both active and passive sources at the surface or in the subsurface can be used to reconstruct a valid model of the subsurface. Seismic interferometry can produce a result similar to traditional methods without limitations on the diffusivity of the wavefield or ambient sources. In a drilling application, it is possible to utilize a virtual source to image the subsurface adjacent to a downhole location. This application is increasingly utilized particularly for exploration in subsalt settings.

Mathematical and Physical Explanation

Seismic interferometry provides for the possibility of reconstructing the subsurface reflection response using the crosscorrelations of two seismic traces. Recent work has mathematically demonstrated applications of crosscorrelation for reconstructing Green's function using wave field reci-procity theorem in a lossless, 3D heterogeneous medium. Traces are most often extended records of passive background noise, but it is also possible to utilize active sources depending on the objective. Seismic interferometry essentially exploits the phase difference between adjacent receiver locations to image the subsurface.

Seismic interferometry consists of simple crosscorrelation and stacking of actual receiver responses to approximate the impulse response as if a virtual source was placed at the location of the applicable receiver. Crosscorrelation of continuous functions in the time domain is:

$$(f_1 * f_2)(t) = \int f_1(\lambda) f_2(\lambda - t) d\lambda$$

Where the functions are integrated as a function of time at different lag values. In fact, crosscorrelation can be understood conceptually as the traveltime lag associated with waveforms in two discrete receiver locations. Crosscorrelation is similar to convolution where the second function is folded relative to the first.

Seismic interferometry is fundamentally similar to the optical interferogram produced by the interference of a direct and reflected wave passing through a glass lens where intensity is primarily dependent upon the phase component.

$$I = 1 + 2R \cdot 2\cos[\omega(\lambda Ar + \lambda rB)] + R^4$$

Where, intensity is related to the magnitude of the reflection coefficient (R) and the phase component $\omega(\lambda Ar + \lambda rB)$. An estimate of the reflectivity distributions can be obtained through the crosscorrelation of the direct wave at a location A with the reflection recorded at a location B where A represents the reference trace. The multiplication of the conjugate of the trace spectrum at A and the trace spectrum at B gives:

$$\Phi_{AB} = Re^{i\omega(\lambda Ar + \lambda rB)} + o.t.$$

Where: Φ_{AB} = product spectrum o.t. = additional terms, e.g. correlations of direct-direct, etc. As in the previous case, the product spectrum is a function of phase.

Changes in reflector geometry lead to changes in the correlation result and the reflector geometry can be recovered through the application of a migration kernel. Interpretation of raw interferograms is not normally attempted; crosscorrelated results are generally processed using some form of migration.

In the simplest case, consider a rotating drill bit at depth radiating energy that is recorded by geophones on the surface. It is possible to assume that the phase of the source wavelet at a given position is random and utilize the crosscorrelation of the direct wave at a location A with a ghost reflection at a location B to image a subsurface reflector without any knowledge regarding the source location. The crosscorrelation of traces A and B in the frequency domain simplifies as:

$$\Phi(A,B) = -(Wi\omega)^2 Re^{i\omega(\lambda Ar\lambda rB)} + o.t.$$

Where, $Wi(\omega)$ = frequency domain source wavelet (ith wavelet).

The crosscorrelation of the direct wave at a location A with a ghost reflection at a location B removes the unknown source term where:

$$\Phi(A,B) \approx Re^{i\omega(\lambda Ar\lambda rB)}$$

This form is equivalent to a virtual source configuration at a location A imaging hypothetical reflections at a location B. Migration of these correlation positions removes the phase term and yields a final migration image at position x where:

$$m(x) = \Sigma\phi(A,B,\lambda Ax+\lambda xB)$$

Where, $\phi(A,B,t)$ = temporal correlation between locations A and B with lag time t.

This model has been applied to simulate subsurface geometry in West Texas using simulated models including a traditional buried source and a synthetic (virtual) rotating drill bit source to produce similar results. A similar model demonstrated the reconstruction of a simulated subsurface geometry. In this case, the reconstructed subsurface response correctly modeled the relative positions of primaries and multiples. Additional equations can be derived to reconstruct signal geometries in a wide variety of cases.

Applications

Seismic interferometry is currently utilized primarily in research and academic settings. In one example, passive listening and the crosscorrelation of long noise traces was used to approximate the impulse response for shallow subsurface velocity analysis in Southern California. Seismic interferometry provided a result comparable to that indicated using elaborate inversion techniques. Seismic interferometry is most often used for the examination of the near surface and is often utilized to reconstruct surface and direct waves only. As such, seismic interferometry is commonly used to estimate ground roll to aid in its removal. Seismic interferometry simplifies estimates of shear wave velocity and attenuation in a standing building. Seismic interferometry has been applied to image the seismic scattering and velocity structure of volcanoes.

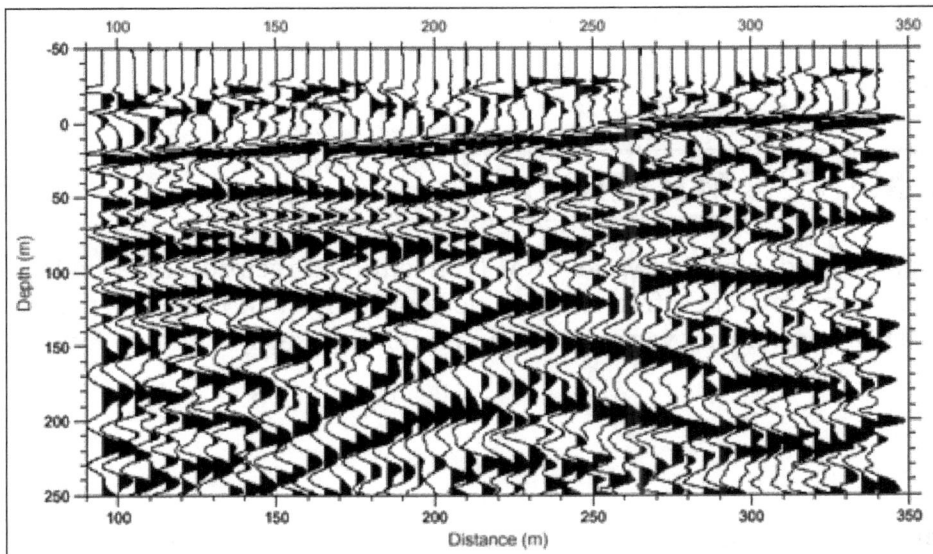

Seismic display as recorded by geophones.

Exploration and Production

Increasingly, seismic interferometry is finding a place in exploration and production. SI can image dipping sediments adjacent to salt domes. Complex salt geometries are poorly resolved using traditional seismic reflection techniques. An alternative method calls for the use of downhole sources and receivers adjacent to subsurface salt features. It is often difficult to generate an ideal seismic signal in a downhole location. Seismic interferometry can virtually move a source into a downhole location to better illuminate and capture steeply dipping sediments on the flank of a salt dome. In this case, the SI result was very similar to that obtained using an actual downhole source. Seismic interferometry can locate the position of an unknown source and is often utilized in hydrofrac applications to map the extent of induced fractures. It is possible that interferometric techniques can be applied to timelapse seismic monitoring of subtle changes in reservoir properties in the subsurface.

Limitations

Seismic interferometry applications are currently limited by a number of factors. Real world media and noise represent limitations for current theoretical development. For example, for interferometry to work noise sources must be uncorrelated and completely surround the region of interest. In addition, attenuation and geometrical spreading are largely neglected and need to be incorporated into more robust models. Other challenges are inherent to seismic interferometry. For example, the source term only drops out in the case of the crosscorrelation of a direct wave at a location A with a ghost reflection at a location B. The correlation of other waveforms can introduce multiples to the resulting interferogram. Velocity analysis and filtering can reduce but not eliminate the occurrence of multiples in a given dataset.

Although there have been many advancements in seismic interferometry challenges still remain. One of the biggest remaining challenges is extending the theory to account for real world media and noise distributions in the subsurface. Natural sources typically do not comply with mathematical generalizations and may in fact display some degree of correlation. Additional problems must be addressed before applications of seismic interferometry can become more widespread.

PLUS-MINUS METHOD

The plus-minus method, also known as CRM (conventional reciprocal method), is a geophysical method to analyze seismic refraction data developed by J. G. Hagedoorn. It can be used to calculate the depth and velocity variations of an undulating layer boundary for slope angles less than ~10°.

Theory

In the plus-minus method, the near surface is modeled as a layer above a halfspace where both the layer and the halfspace are allowed to have varying velocities. The method is based on the analysis of the so-called 'plus time' t^+ and 'minus time' t^- that are given by:

$$t^+ = t_{AX} + t_{BX} - t_{AB}$$

$$t^- = t_{AX} - t_{BX} - t_{AB}$$

where t_{AB} is the traveltime from A to B, t_{AX} the traveltime from A to X and t_{BX} the traveltime from B to X.

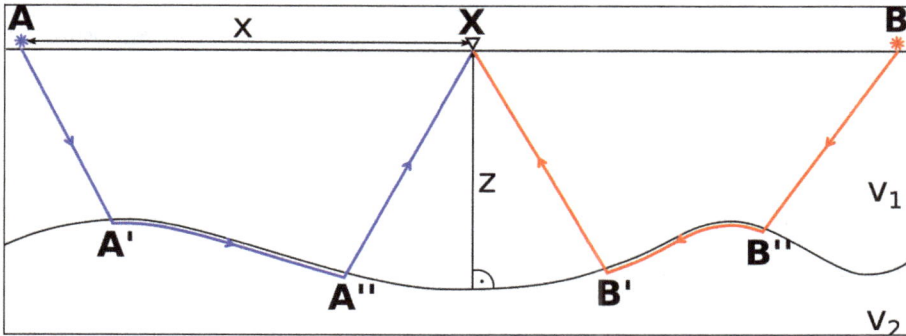

Sketch of the plus-minus method illustrating the ray paths between sources A and B and the receiver X.

Assuming that the layer boundary is planar between A" and B" and that the dip is small (<10°), the plus time t^+ corresponds to the intercept time in classic refraction analysis and the minus time t^- can be expressed as:

$$t^- = t^+ + \frac{2x}{v_2}$$

where x is the offset between A and X and v_2 is the velocity of the halfspace.

Therefore, the slope of the minus time $\Delta t^- / \Delta x$ can be used to estimate the velocity of the halfspace v_2:

$$v_2(x) = 2 \frac{\Delta x}{\Delta t^-}$$

The interval Δx over which the slope is estimated should be chosen according to data quality. A larger Δx results in more stable velocity estimates but also introduces stronger smoothing. Like in classical refraction analysis, the thickness of the upper layer can be derived from the intercept time t^+:

$$z(x) = \frac{t^+ v_1(x) v_2(x)}{2\sqrt{v_2^2 - v_1^2}}$$

This requires an estimation of the velocity of the upper layer $v_1(x)$ which can be obtained from the direct wave in the traveltime diagram.

Furthermore, the results of the plus-minus method can be used to calculate the shot-receiver static shift $\Delta\tau(x)$:

$$\Delta\tau(x) = -\frac{z(x)}{v_1(x)} + \frac{E_X - E_S + z(x)}{v_2(x)}$$

where E_X is the datum elevation and E_S the surface elevation at station X.

Applications

The Plus-minus was developed for shallow seismic surveys where a thin, low velocity weathering layer covers the more solid basement. The thickness of the weathering layer is, among others, important for static corrections in reflection seismic processing or for engineering purposes. An important advantage of the method is that it does not require manual interpretation of the intercept time or the crossover point. This makes it is also easy to implement in computer programs. However, it is only applicable if the layer boundary is planar in parts and the dips are small. These assumptions often lead to smoothing of the actual topography of the layer boundary. Nowadays, the Plus-minus method has mostly been replaced by more advanced inversion methods that have less restrictions. However, the Plus-minus method is still used for real-time processing in the field because of its simplicity and low computational costs.

ACCELEROGRAPH

An accelerograph is recorder that uses an accelerometer, which as you can tell from the name, detects the acceleration of the ground. Accelerometers are much less sensitive than seismometers, but have a much greater range, detecting ±2g or more of ground acceleration (things start flying off the ground at 1g, when gravity is overcome). By comparison a seismometer will clip at full scale if you tap it too hard with your finger.

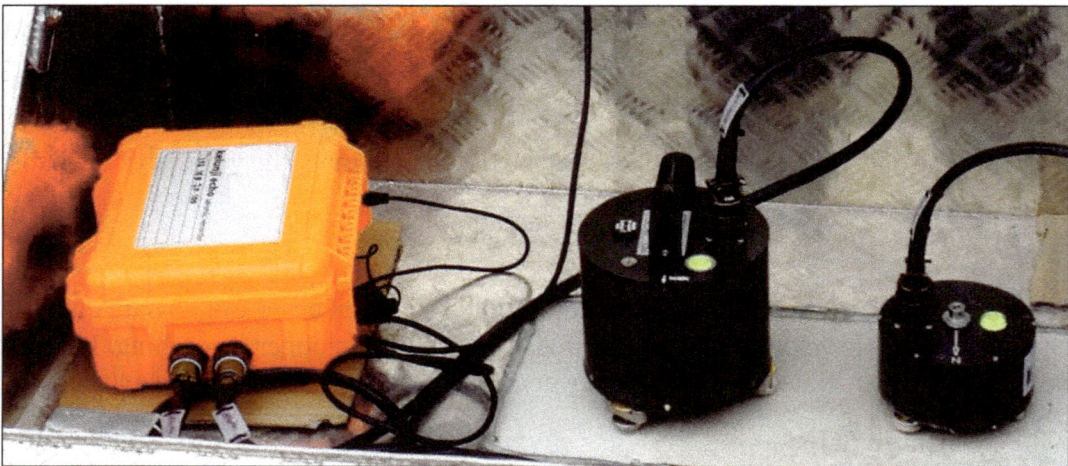

So, seismometers are good for detecting very small levels of ground motion (from very small or very distant events), and accelerometers are good at recording strong ground motion that is potentially damaging at the recording location. We will often install an earthquake recording station using both types of sensor to get the best of both worlds.

SMAs, or strong motion accelerographs, are usually all that is required for monitoring the response of a structure during an earthquake, whether this is a building, bridge, dam, power station, or any other critical infrastructure that could be affected by a large earthquake. Signals that are too weak to be clearly visible on an accelerograph will generally not be of any concern to the structural integrity of the asset.

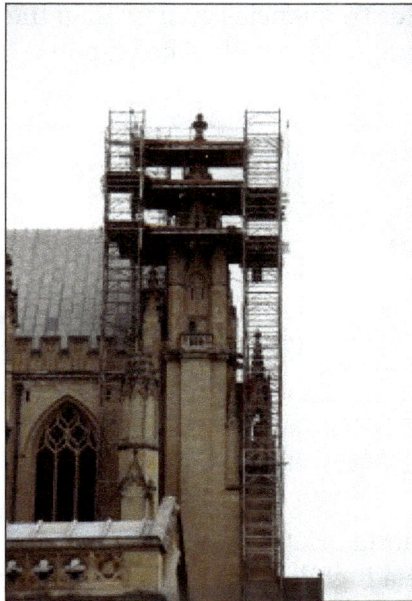

The SRC has used various types of SMAs over the years. Those with MEMS accelerometers use tiny sensors like those found in cars to trigger airbags on impact, but are much more sensitive. They are less sensitive than dedicated earthquake accelerometers, but are still sufficient to record acceleration from significant earthquakes as the lower sensitivity limit is still well below damage thresholds. For monitoring the natural frequency and modal response of structures to earthquakes, more sensitive earthquake accelerometers are required.

GEOPHONE

A geophone is a ground motion transducer that has been used by geophysicists and seismologists to convert ground movement into voltage. Any deviation in this measured voltage from the base line is regarded as seismic response, which is used for analyzing the earth's structure.

Resonance frequency is the key factor in a geophone, and it has to be low for the measurement of low-frequency signals. On the other hand, geophone must exhibit high bandwidth to measure high-frequency signals as well. However, most of the currently available geophones include mechanical springs that decrease the performance of the device.

Feeding back the output of geophone can vary the geophone's sensitivity with respect to its frequency, and as a result low frequency signals are amplified. The resolution of geophone can be enhanced by controlling the position of proof mass.

Working Principle

A typical geophone consists of a mass suspended by means of mechanical springs. The geophone housing and the suspended mass start moving with the application of a velocity at frequencies lesser than the resonance frequency.

The mass will remain stationary for frequencies greater than the resonance frequency. The movement of mass is based on either magnets or coils. The response of a coil/magnet geophone is proportional to the ground velocity.

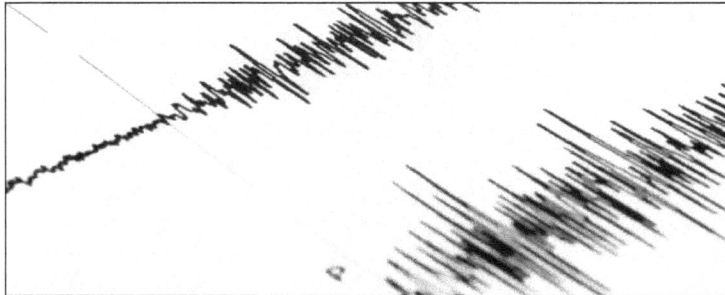

Applications

Geophones are used for several industrial applications for vibration isolation purposes and absolute velocity sensing to achieve a high level of accuracy and precision. Geophone technology is employed to measure absolute velocity for lithographic and high-level inspection applications to determine payload disturbances caused by moving parts and other external disturbances.

They are also used to position and control a complex lens system. Detection of leakage in oil and gas fields and earthquake prediction are also other major applications of geophones.

References

- Dondurur, Derman; Karslı, Hakan (May 2012). "Swell Noise Suppression by Wiener Prediction Filter". Journal of Applied Geophysics. 80: 91–100. doi:10.1016/j.jappgeo.2012.02.001

- Varga, P. (2008). "History of Early Isoseismal Maps". Acta Geodaetica et Geophysica Hungarica. 43 (2–3): 285–307. doi:10.1556/AGeod.43.2008.2-3.15

- Musson, R.M.W. (2000). "Intensity-based seismic risk assessment". Soil Dynamics and Earthquake Engineering. 20 (5–8): 353–360. doi:10.1016/s0267-7261(00)00083-x

- Draganov, D.; Wapenaar, K.; Thorbecke, J. (2006). "Seismic interferometry: Reconstructing the earth's reflection response". Geophysics. 71 (4): SI61–SI70. Bibcode:2006Geop...71...61D. CiteSeerX 10.1.1.75.113. doi:10.1190/1.2209947

- Strong-motion-accelerographs: src.com.au, Retrieved 25 July, 2019

- Overmeeren, R. A. (2001). "Hagedoorn's plus-minus method: the beauty of simplicity". Geophysical Prospecting. 49 (6): 687–696. Bibcode:1964GeopP..12....1U. doi:10.1111/j.1365-2478.1964.tb01888.x

Earthquakes: An Integrated Study

Earthquake is the shaking of the Earth's surface due to the passage of seismic waves through Earth's rocks. Slow earthquake, blind thrust earthquake, megathrust earthquake, submarine earthquake, doublet earthquake, etc. are a few of its types. This chapter discusses these types of earthquakes and related aspects in detail.

Earthquake is any sudden shaking of the ground caused by the passage of seismic waves through Earth's rocks. Seismic waves are produced when some form of energy stored in Earth's crust is suddenly released, usually when masses of rock straining against one another suddenly fracture and "slip." Earthquakes occur most often along geologic faults, narrow zones where rock masses move in relation to one another. The major fault lines of the world are located at the fringes of the huge tectonic plates that make up Earth's crust.

Residents of an earthquake-damaged neighbourhood of Port-au-Prince, Haiti, seeking safety in a sports field. The magnitude-7.0 earthquake struck the region the day before.

Little was understood about earthquakes until the emergence of seismology at the beginning of the 20th century. Seismology, which involves the scientific study of all aspects of earthquakes, has yielded answers to such long-standing questions as why and how earthquakes occur.

About 50,000 earthquakes large enough to be noticed without the aid of instruments occur annually over the entire Earth. Of these, approximately 100 are of sufficient size to produce substantial damage if their centres are near areas of habitation. Very great earthquakes occur on average about once per year. Over the centuries they have been responsible for millions of deaths and an incalculable amount of damage to property.

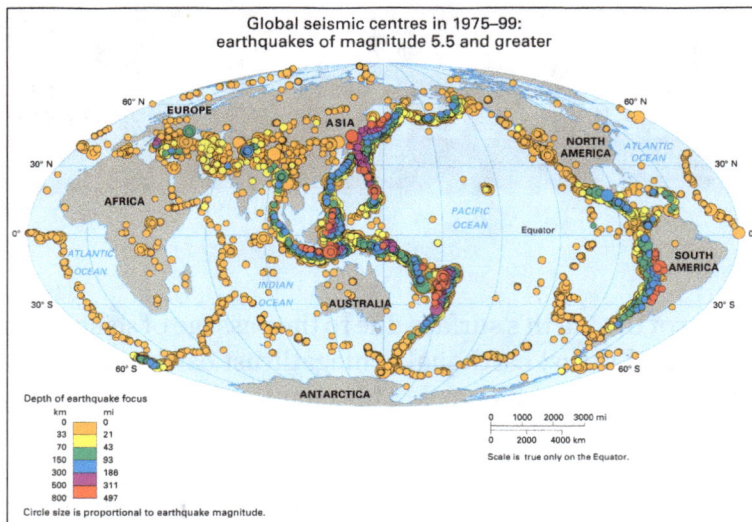

Global seismic centres in 1975–99:
earthquakes of magnitude 5.5 and greater

Crowds watching the fires set off by the earthquake.

Nature of Earthquakes

Causes of Earthquakes

Earth's major earthquakes occur mainly in belts coinciding with the margins of tectonic plates. This has long been apparent from early catalogs of felt earthquakes and is even more readily discernible in modern seismicity maps, which show instrumentally determined epicentres. The most important earthquake belt is the Circum-Pacific Belt, which affects many populated coastal regions around the Pacific Ocean—for example, those of New Zealand, New Guinea, Japan, the Aleutian Islands, Alaska, and the western coasts of North and South America. It is estimated that 80 percent of the energy presently released in earthquakes comes from those whose epicentres are in this belt. The seismic activity is by no means uniform throughout the belt, and there are a number of branches at various points. Because at many places the Circum-Pacific Belt is associated with volcanic activity, it has been popularly dubbed the "Pacific Ring of Fire".

A second belt, known as the Alpide Belt, passes through the Mediterranean region eastward through Asia and joins the Circum-Pacific Belt in the East Indies. The energy released in earthquakes from this belt is about 15 percent of the world total. There also are striking connected belts of seismic activity, mainly along oceanic ridges—including those in the Arctic Ocean, the Atlantic Ocean, and the western Indian Ocean—and along the rift valleys of East Africa. This global seismicity distribution is best understood in terms of its plate tectonic setting.

Natural Forces

Earthquakes are caused by the sudden release of energy within some limited region of the rocks of the Earth. The energy can be released by elastic strain, gravity, chemical reactions, or even the motion of massive bodies. Of all these the release of elastic strain is the most important cause, because this form of energy is the only kind that can be stored in sufficient quantity in the Earth to produce major disturbances. Earthquakes associated with this type of energy release are called tectonic earthquakes.

Tectonics

Tectonic earthquakes are explained by the so-called elastic rebound theory, formulated by the American geologist Harry Fielding Reid after the San Andreas Fault ruptured in 1906, generating the great San Francisco earthquake. According to the theory, a tectonic earthquake occurs when strains in rock masses have accumulated to a point where the resulting stresses exceed the strength of the rocks, and sudden fracturing results. The fractures propagate rapidly through the rock, usually tending in the same direction and sometimes extending many kilometres along a local zone of weakness. In 1906, for instance, the San Andreas Fault slipped along a plane 430 km (270 miles) long. Along this line the ground was displaced horizontally as much as 6 metres (20 feet).

As a fault rupture progresses along or up the fault, rock masses are flung in opposite directions and thus spring back to a position where there is less strain. At any one point this movement may take place not at once but rather in irregular steps; these sudden slowings and restartings give rise to the vibrations that propagate as seismic waves. Such irregular properties of fault rupture are now included in the modeling of earthquake sources, both physically and mathematically. Roughnesses along the fault are referred to as asperities, and places where the rupture slows or stops are said to be fault barriers. Fault rupture starts at the earthquake focus, a spot that in many cases is close to 5–15 km under the surface. The rupture propagates in one or both directions over the fault plane until stopped or slowed at a barrier. Sometimes, instead of being stopped at the barrier, the fault rupture recommences on the far side; at other times the stresses in the rocks break the barrier, and the rupture continues.

Earthquakes have different properties depending on the type of fault slip that causes them. The usual fault model has a "strike" (that is, the direction from north taken by a horizontal line in the fault plane) and a "dip" (the angle from the horizontal shown by the steepest slope in the fault). The lower wall of an inclined fault is called the footwall. Lying over the footwall is the hanging wall. When rock masses slip past each other parallel to the strike, the movement is known as strike-slip faulting. Movement parallel to the dip is called dip-slip faulting. Strike-slip faults are right lateral or left lateral, depending on whether the block on the opposite side of the fault from an observer has moved to the right or left. In dip-slip faults, if the hanging-wall block moves downward relative to the footwall block, it is called "normal" faulting; the opposite motion, with the hanging wall moving upward relative to the footwall, produces reverse or thrust faulting.

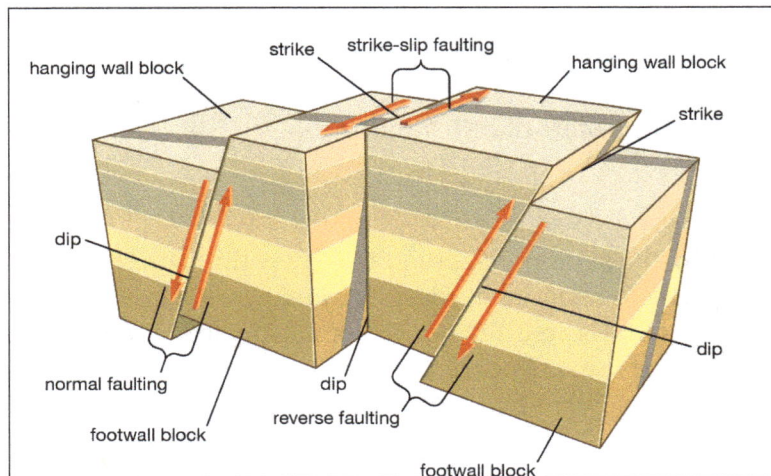

Types of faulting in tectonic earthquakes. In normal and reverse faulting, rock masses slip vertically past each other. In strike-slip faulting, the rocks slip past each other horizontally.

All known faults are assumed to have been the seat of one or more earthquakes in the past, though tectonic movements along faults are often slow, and most geologically ancient faults are now aseismic (that is, they no longer cause earthquakes). The actual faulting associated with an earthquake may be complex, and it is often not clear whether in a particular earthquake the total energy issues from a single fault plane.

Observed geologic faults sometimes show relative displacements on the order of hundreds of kilometres over geologic time, whereas the sudden slip offsets that produce seismic waves may range from only several centimetres to tens of metres. In the 1976 Tangshan earthquake, for example, a surface strike-slip of about one metre was observed along the causative fault east of Beijing, and in the 1999 Taiwan earthquake the Chelung-pu fault slipped up to eight metres vertically.

Volcanism

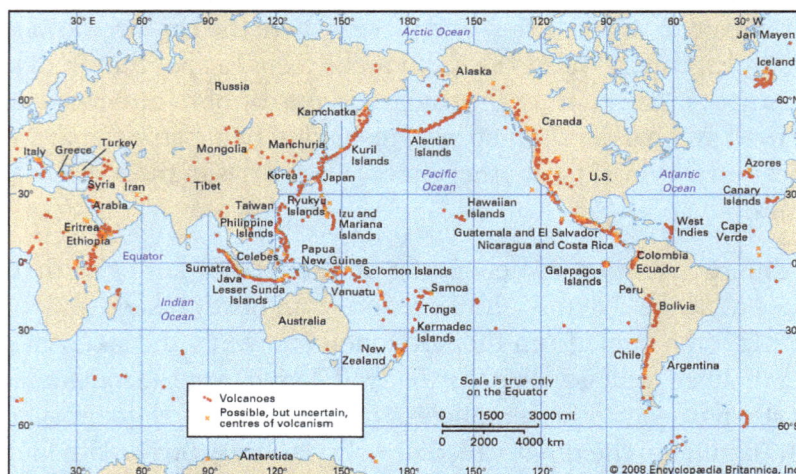

Volcanoes and thermal fields that have been active during the past 10,000 years.

A separate type of earthquake is associated with volcanic activity and is called a volcanic earthquake. Yet it is likely that even in such cases the disturbance is the result of a sudden slip of rock masses adjacent to the volcano and the consequent release of elastic strain energy. The stored energy, however, may in part be of hydrodynamic origin due to heat provided by magma moving in reservoirs beneath the volcano or to the release of gas under pressure.

There is a clear correspondence between the geographic distribution of volcanoes and major earthquakes, particularly in the Circum-Pacific Belt and along oceanic ridges. Volcanic vents, however, are generally several hundred kilometres from the epicentres of most major shallow earthquakes, and many earthquake sources occur nowhere near active volcanoes. Even in cases where an earthquake's focus occurs directly below structures marked by volcanic vents, there is probably no immediate causal connection between the two activities; most likely both are the result of the same tectonic processes.

Artificial Induction

Earthquakes are sometimes caused by human activities, including the injection of fluids into deep wells, the detonation of large underground nuclear explosions, the excavation of mines, and the filling of large reservoirs. In the case of deep mining, the removal of rock produces changes in the strain around the tunnels. Slip on adjacent, preexisting faults or outward shattering of rock into the new cavities may occur. In fluid injection, the slip is thought to be induced by premature release of elastic strain, as in the case of tectonic earthquakes, after fault surfaces are lubricated by the liquid. Large underground nuclear explosions have been known to produce slip on already strained faults in the vicinity of the test devices.

Reservoir Induction

Of the various earthquake-causing activities cited above, the filling of large reservoirs is among the most important. More than 20 significant cases have been documented in which local seismicity has increased following the impounding of water behind high dams. Often, causality cannot be substantiated, because no data exists to allow comparison of earthquake occurrence before and after the reservoir was filled. Reservoir-induction effects are most marked for reservoirs exceeding 100 metres (330 feet) in depth and 1 cubic km (0.24 cubic mile) in volume. Three sites where such connections have very probably occurred are the Hoover Dam in the United States, the Aswan High Dam in Egypt, and the Kariba Dam on the border between Zimbabwe and Zambia. The most generally accepted explanation for earthquake occurrence in such cases assumes that rocks near the reservoir are already strained from regional tectonic forces to a point where nearby faults are almost ready to slip. Water in the reservoir adds a pressure perturbation that triggers the fault rupture. The pressure effect is perhaps enhanced by the fact that the rocks along the fault have lower strength because of increased water-pore pressure. These factors notwithstanding, the filling of most large reservoirs has not produced earthquakes large enough to be a hazard.

The specific seismic source mechanisms associated with reservoir induction have been established in a few cases. For the main shock at the Koyna Dam and Reservoir in India, the evidence favours strike-slip faulting motion. At both the Kremasta Dam in Greece and the Kariba Dam in Zimbabwe-Zambia, the generating mechanism was dip-slip on normal faults. By contrast, thrust mechanisms have been determined for sources of earthquakes at the lake behind Nurek Dam in Tajikistan. More than 1,800 earthquakes occurred during the first nine years after water was impounded in this 317-metre-deep reservoir in 1972, a rate amounting to four times the average number of shocks in the region prior to filling.

Seismology and Nuclear Explosions

In 1958 representatives from several countries, including the United States and the Soviet Union, met to discuss the technical basis for a nuclear test-ban treaty. Among the matters considered was the feasibility of developing effective means with which to detect underground nuclear explosions and to distinguish them seismically from earthquakes. After that conference, much special research was directed to seismology, leading to major advances in seismic signal detection and analysis.

Recent seismological work on treaty verification has involved using high-resolution seismographs in a worldwide network, estimating the yield of explosions, studying wave attenuation in the Earth, determining wave amplitude and frequency spectra discriminants, and applying seismic arrays. The findings of such research have shown that underground nuclear explosions, compared with natural earthquakes, usually generate seismic waves through the body of the Earth that are of much larger amplitude than the surface waves. This telltale difference along with other types of seismic evidence suggest that an international monitoring network of 270 seismographic stations could detect and locate all seismic events over the globe of magnitude 4 and above (corresponding to an explosive yield of about 100 tons of TNT).

Effects of Earthquakes

Earthquakes have varied effects, including changes in geologic features, damage to man-made structures, and impact on human and animal life. Most of these effects occur on solid ground, but, since most earthquake foci are actually located under the ocean bottom, severe effects are often observed along the margins of oceans.

Surface Phenomena

Earthquakes often cause dramatic geomorphological changes, including ground movements—either vertical or horizontal—along geologic fault traces; rising, dropping, and tilting of the ground surface; changes in the flow of groundwater; liquefaction of sandy ground; landslides; and mudflows. The investigation of topographic changes is aided by geodetic measurements, which are made systematically in a number of countries seriously affected by earthquakes.

Earthquakes can do significant damage to buildings, bridges, pipelines, railways, embankments, and other structures. The type and extent of damage inflicted are related to the strength of the ground motions and to the behaviour of the foundation soils. In the most intensely damaged region, called the meizoseismal area, the effects of a severe earthquake are usually complicated and depend on the topography and the nature of the surface materials. They are often more severe on soft alluvium and unconsolidated sediments than on hard rock. At distances of more than 100 km (60 miles) from the source, the main damage is caused by seismic waves traveling along the surface. In mines there is frequently little damage below depths of a few hundred metres even though the ground surface immediately above is considerably affected.

Earthquakes are frequently associated with reports of distinctive sounds and lights. The sounds are generally low-pitched and have been likened to the noise of an underground train passing through a station. The occurrence of such sounds is consistent with the passage of high-frequency seismic waves through the ground. Occasionally, luminous flashes, streamers, and bright balls have been reported in the night sky during earthquakes. These lights have been attributed to electric induction in the air along the earthquake source.

Tsunamis

Following certain earthquakes, very long-wavelength water waves in oceans or seas sweep inshore. More properly called seismic sea waves or tsunamis (tsunami is a Japanese word for "harbour wave"), they are commonly referred to as tidal waves, although the attractions of the Moon and Sun play no role in their formation. They sometimes come ashore to great heights—tens of metres above mean tide level—and may be extremely destructive.

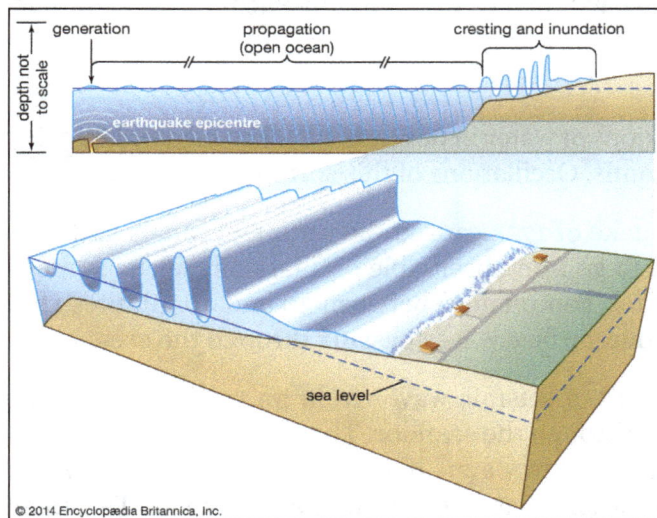

After being generated by an undersea earthquake or landslide, a tsunami may propagate unnoticed over vast reaches of open ocean before cresting in shallow water and inundating a coastline.

The usual immediate cause of a tsunami is sudden displacement in a seabed sufficient to cause the sudden raising or lowering of a large body of water. This deformation may be the fault source of an earthquake, or it may be a submarine landslide arising from an earthquake. Large volcanic eruptions along shorelines, such as those of Thera and Krakatoa, have also produced notable tsunamis. The most destructive tsunami ever recorded occurred on December 26, 2004, after an earthquake displaced the seabed off the coast of Sumatra, Indonesia. More than 200,000 people were killed by a series of waves that flooded coasts from Indonesia to Sri Lanka and even washed ashore on the Horn of Africa.

Following the initial disturbance to the sea surface, water waves spread in all directions. Their speed of travel in deep water is given by the formula $(\sqrt{gh})$, where h is the sea depth and g is the acceleration of gravity. This speed may be considerable—100 metres per second (225 miles per hour) when h is 1,000 metres (3,300 feet). However, the amplitude (that is, the height of disturbance) at the water surface does not exceed a few metres in deep water, and the principal wavelength may be on the order of hundreds of kilometres; correspondingly, the principal wave period—that is, the time interval between arrivals of successive crests—may be on the order of tens of minutes. Because of these features, tsunami waves are not noticed by ships far out at sea.

When tsunamis approach shallow water, however, the wave amplitude increases. The waves may occasionally reach a height of 20 to 30 metres above mean sea level in U- and V-shaped harbours and inlets. They characteristically do a great deal of damage in low-lying ground around such inlets. Frequently, the wave front in the inlet is nearly vertical, as in a tidal bore, and the speed of onrush may be on the order of 10 metres per second. In some cases there are several great waves separated by intervals of several minutes or more. The first of these waves is often preceded by an extraordinary recession of water from the shore, which may commence several minutes or even half an hour beforehand.

Organizations, notably in Japan, Siberia, Alaska, and Hawaii, have been set up to provide tsunami warnings. A key development is the Seismic Sea Wave Warning System, an internationally supported system designed to reduce loss of life in the Pacific Ocean. Centred in Honolulu, it issues alerts based on reports of earthquakes from circum-Pacific seismographic stations.

Seiches

Seiches are rhythmic motions of water in nearly landlocked bays or lakes that are sometimes induced by earthquakes and tsunamis. Oscillations of this sort may last for hours or even for a day or two.

The great Lisbon earthquake of 1755 caused the waters of canals and lakes in regions as far away as Scotland and Sweden to go into observable oscillations. Seiche surges in lakes in Texas, in the southwestern United States, commenced between 30 and 40 minutes after the 1964 Alaska earthquake, produced by seismic surface waves passing through the area.

A related effect is the result of seismic waves from an earthquake passing through the seawater following their refraction through the seafloor. The speed of these waves is about 1.5 km (0.9 mile) per second, the speed of sound in water. If such waves meet a ship with sufficient intensity, they give the impression that the ship has struck a submerged object. This phenomenon is called a seaquake.

Intensity and Magnitude of Earthquakes

Intensity Scales

The violence of seismic shaking varies considerably over a single affected area. Because the entire range of observed effects is not capable of simple quantitative definition, the strength of the shaking is commonly estimated by reference to intensity scales that describe the effects in qualitative terms. Intensity scales date from the late 19th and early 20th centuries, before seismographs capable of accurate measurement of ground motion were developed. Since that time, the divisions in these scales have been associated with measurable accelerations of the local ground shaking. Intensity depends, however, in a complicated way not only on ground accelerations but also on the periods and other features of seismic waves, the distance of the measuring point from the source, and the local geologic structure. Furthermore, earthquake intensity, or strength, is distinct from earthquake magnitude, which is a measure of the amplitude, or size, of seismic waves as specified by a seismograph reading.

A number of different intensity scales have been set up during the past century and applied to both current and ancient destructive earthquakes. For many years the most widely used was a 10-point scale devised in 1878 by Michele Stefano de Rossi and François-Alphonse Forel. The scale now generally employed in North America is the Mercalli scale, as modified by Harry O. Wood and Frank Neumann in 1931, in which intensity is considered to be more suitably graded. A 12-point abridged form of the modified Mercalli scale is provided below. Modified Mercalli intensity VIII is roughly correlated with peak accelerations of about one-quarter that of gravity (g = 9.8 metres, or 32.2 feet, per second squared) and ground velocities of 20 cm (8 inches) per second. Alternative scales have been developed in both Japan and Europe for local conditions. The European (MSK) scale of 12 grades is similar to the abridged version of the Mercalli.

Modified Mercalli Scale of Earthquake Intensity

- Not felt. Marginal and long-period effects of large earthquakes.

- Felt by persons at rest, on upper floors, or otherwise favourably placed to sense tremors.

- Felt indoors. Hanging objects swing. Vibrations are similar to those caused by the passing of light trucks. Duration can be estimated.

- Vibrations are similar to those caused by the passing of heavy trucks (or a jolt similar to that caused by a heavy ball striking the walls). Standing automobiles rock. Windows, dishes, doors rattle. Glasses clink, crockery clashes. In the upper range of grade IV, wooden walls and frames creak.

- Felt outdoors; direction may be estimated. Sleepers awaken. Liquids are disturbed, some spilled. Small objects are displaced or upset. Doors swing, open, close. Pendulum clocks stop, start, change rate.

- Felt by all; many are frightened and run outdoors. Persons walk unsteadily. Pictures fall off walls. Furniture moves or overturns. Weak plaster and masonry cracks. Small bells ring (church, school). Trees, bushes shake.

- Difficult to stand. Noticed by drivers of automobiles. Hanging objects quivering. Furniture broken. Damage to weak masonry. Weak chimneys broken at roof line. Fall of plaster, loose bricks, stones, tiles, cornices. Waves on ponds; water turbid with mud. Small slides and caving along sand or gravel banks. Large bells ringing. Concrete irrigation ditches damaged.

- Steering of automobiles affected. Damage to masonry; partial collapse. Some damage to reinforced masonry; none to reinforced masonry designed to resist lateral forces. Fall of stucco and some masonry walls. Twisting, fall of chimneys, factory stacks, monuments, towers, elevated tanks. Frame houses moved on foundations if not bolted down; loose panel walls thrown out. Decayed pilings broken off. Branches broken from trees. Changes in flow or temperature of springs and wells. Cracks in wet ground and on steep slopes.

- General panic. Weak masonry destroyed; ordinary masonry heavily damaged, sometimes with complete collapse; reinforced masonry seriously damaged. Serious damage to reservoirs. Underground pipes broken. Conspicuous cracks in ground. In alluvial areas, sand and mud ejected; earthquake fountains, sand craters.

- Most masonry and frame structures destroyed with their foundations. Some well-built wooden structures and bridges destroyed. Serious damage to dams, dikes, embankments. Large landslides. Water thrown on banks of canals, rivers, lakes, and so on. Sand and mud shifted horizontally on beaches and flat land. Railway rails bent slightly.

- Rails bent greatly. Underground pipelines completely out of service.

- Damage nearly total. Large rock masses displaced. Lines of sight and level distorted. Objects thrown into air.

With the use of an intensity scale, it is possible to summarize such data for an earthquake by constructing isoseismal curves, which are lines that connect points of equal intensity. If there were complete symmetry about the vertical through the earthquake's focus, isoseismals would be circles with the epicentre (the point at the surface of the Earth immediately above where the earthquake originated) as the centre. However, because of the many unsymmetrical geologic factors influencing intensity, the curves are often far from circular. The most probable position of the epicentre is often assumed to be at a point inside the area of highest intensity. In some cases, instrumental data verify this calculation, but not infrequently the true epicentre lies outside the area of greatest intensity.

Earthquake Magnitude

Earthquake magnitude is a measure of the "size," or amplitude, of the seismic waves generated by an earthquake source and recorded by seismographs. Because the size of earthquakes varies enormously, it is necessary for purposes of comparison to compress the range of wave amplitudes measured on seismograms by means of a mathematical device. In 1935 the American seismologist Charles F. Richter set up a magnitude scale of earthquakes as the logarithm to base 10 of the maximum seismic wave amplitude (in thousandths of a millimetre) recorded on a standard seismograph (the Wood-Anderson torsion pendulum seismograph) at a distance of 100 km (60 miles) from the earthquake epicentre. Reduction of amplitudes observed at various distances to the amplitudes expected at the standard distance of 100 km is made on the basis of empirical tables. Richter magnitudes ML are computed on the assumption that the ratio of the maximum wave amplitudes at two given distances is the same for all earthquakes and is independent of azimuth.

Richter first applied his magnitude scale to shallow-focus earthquakes recorded within 600 km of the epicentre in the southern California region. Later, additional empirical tables were set up, whereby observations made at distant stations and on seismographs other than the standard type could be used. Empirical tables were extended to cover earthquakes of all significant focal depths and to enable independent magnitude estimates to be made from body- and surface-wave observations. A current form of the Richter scale is shown in the table.

Richter scale of earthquake magnitude			
Magnitude level	Category	Effects	Earthquakes per year
Less than 1.0 to 2.9	Micro	Generally not felt by people, though recorded on local instruments.	More than 100,000
3.0–3.9	Minor	Felt by many people; no damage.	12,000–100,000

4.0–4.9	Light	Felt by all; minor breakage of objects.	2,000–12,000
5.0–5.9	Moderate	Some damage to weak structures.	200–2,000
6.0–6.9	Strong	Moderate damage in populated areas.	20–200
7.0–7.9	Major	Serious damage over large areas; loss of life.	3–20
8.0 and higher	Great	Severe destruction and loss of life over large areas.	Fewer than 3

At the present time a number of different magnitude scales are used by scientists and engineers as a measure of the relative size of an earthquake. The P-wave magnitude (M_b), for one, is defined in terms of the amplitude of the P wave recorded on a standard seismograph. Similarly, the surface-wave magnitude (M_s) is defined in terms of the logarithm of the maximum amplitude of ground motion for surface waves with a wave period of 20 seconds.

As defined, an earthquake magnitude scale has no lower or upper limit. Sensitive seismographs can record earthquakes with magnitudes of negative value and have recorded magnitudes up to about 9.0. The 1906 San Francisco earthquake, for example, had a Richter magnitude of 8.25.

A scientific weakness is that there is no direct mechanical basis for magnitude as defined above. Rather, it is an empirical parameter analogous to stellar magnitude assessed by astronomers. In modern practice a more soundly based mechanical measure of earthquake size is used—namely, the seismic moment (M_o). Such a parameter is related to the angular leverage of the forces that produce the slip on the causative fault. It can be calculated both from recorded seismic waves and from field measurements of the size of the fault rupture. Consequently, seismic moment provides a more uniform scale of earthquake size based on classical mechanics. This measure allows a more scientific magnitude to be used called moment magnitude (M_w). It is proportional to the logarithm of the seismic moment; values do not differ greatly from M_s values for moderate earthquakes. Given the above definitions, the great Alaska earthquake of 1964, with a Richter magnitude (M_L) of 8.3, also had the values $M_s = 8.4$, $M_o = 820 \times 10^{27}$ dyne centimetres, and $M_w = 9.2$.

Earthquake Energy

Energy in an earthquake passing a particular surface site can be calculated directly from the recordings of seismic ground motion, given, for example, as ground velocity. Such recordings indicate an energy rate of 10^5 watts per square metre (9,300 watts per square foot) near a moderate-size earthquake source. The total power output of a rupturing fault in a shallow earthquake is on the order of 10^{14} watts, compared with the 10^5 watts generated in rocket motors.

The surface-wave magnitude M_s has also been connected with the surface energy E_s of an earthquake by empirical formulas. These give $E_s = 6.3 \times 10^{11}$ and 1.4×10^{25} ergs for earthquakes of $M_s = 0$ and 8.9, respectively. A unit increase in M_s corresponds to approximately a 32-fold increase in energy. Negative magnitudes M_s correspond to the smallest instrumentally recorded earthquakes, a magnitude of 1.5 to the smallest felt earthquakes, and one of 3.0 to any shock felt at a distance of up to 20 km (12 miles). Earthquakes of magnitude 5.0 cause light damage near the epicentre; those of 6.0 are destructive over a restricted area; and those of 7.5 are at the lower limit of major earthquakes.

The total annual energy released in all earthquakes is about 10^{25} ergs, corresponding to a rate of work between 10 million and 100 million kilowatts. This is approximately one one-thousandth the annual amount of heat escaping from the Earth's interior. Ninety percent of the total seismic energy comes from earthquakes of magnitude 7.0 and higher—that is, those whose energy is on the order of 10^{23} ergs or more.

Frequency

There also are empirical relations for the frequencies of earthquakes of various magnitudes. Suppose N to be the average number of shocks per year for which the magnitude lies in a range about M_s. Then,

$$\log_{10} N = a - bM_s$$

fits the data well both globally and for particular regions; for example, for shallow earthquakes worldwide, a = 6.7 and b = 0.9 when M_s > 6.0. The frequency for larger earthquakes therefore increases by a factor of about 10 when the magnitude is diminished by one unit. The increase in frequency with reduction in M_s falls short, however, of matching the decrease in the energy E. Thus, larger earthquakes are overwhelmingly responsible for most of the total seismic energy release. The number of earthquakes per year with M_b > 4.0 reaches 50,000.

Occurrence of Earthquakes

Tectonic Associations

Global seismicity patterns had no strong theoretical explanation until the dynamic model called plate tectonics was developed during the late 1960s. This theory holds that the Earth's upper shell, or lithosphere, consists of nearly a dozen large, quasi-stable slabs called plates. The thickness of each of these plates is roughly 80 km (50 miles). The plates move horizontally relative to neighbouring plates at a rate of 1 to 10 cm (0.4 to 4 inches) per year over a shell of lesser strength called the asthenosphere. At the plate edges where there is contact between adjoining plates, boundary tectonic forces operate on the rocks, causing physical and chemical changes in them. New lithosphere is created at oceanic ridges by the upwelling and cooling of magma from the Earth's mantle. The horizontally moving plates are believed to be absorbed at the ocean trenches, where a subduction process carries the lithosphere downward into the Earth's interior. The total amount of lithospheric material destroyed at these subduction zones equals that generated at the ridges.

Seismological evidence (such as the location of major earthquake belts) is everywhere in agreement with this tectonic model. Earthquake sources are concentrated along the oceanic ridges, which correspond to divergent plate boundaries. At the subduction zones, which are associated with convergent plate boundaries, intermediate- and deep-focus earthquakes mark the location of the upper part of a dipping lithosphere slab. The focal mechanisms indicate that the stresses are aligned with the dip of the lithosphere underneath the adjacent continent or island arc.

Some earthquakes associated with oceanic ridges are confined to strike-slip faults, called transform faults, that offset the ridge crests. The majority of the earthquakes occurring along such horizontal shear faults are characterized by slip motions. Also in agreement with the plate tectonics theory is the high seismicity encountered along the edges of plates where they slide past each other. Plate boundaries of this kind, sometimes called fracture zones, include the San Andreas Fault in California and the North Anatolian fault system in Turkey. Such plate boundaries are the site of interplate earthquakes of shallow focus.

The low seismicity within plates is consistent with the plate tectonic description. Small to large earthquakes do occur in limited regions well within the boundaries of plates; however, such intraplate seismic events can be explained by tectonic mechanisms other than plate boundary motions and their associated phenomena.

Shallow, Intermediate and Deep Foci

Most parts of the world experience at least occasional shallow earthquakes—those that originate within 60 km (40 miles) of the Earth's outer surface. In fact, the great majority of earthquake foci are shallow. It should be noted, however, that the geographic distribution of smaller earthquakes is less completely determined than more severe quakes, partly because the availability of relevant data is dependent on the distribution of observatories.

Of the total energy released in earthquakes, 12 percent comes from intermediate earthquakes—that is, quakes with a focal depth ranging from about 60 to 300 km. About 3 percent of total energy comes from deeper earthquakes. The frequency of occurrence falls off rapidly with increasing focal depth in the intermediate range. Below intermediate depth the distribution is fairly uniform until the greatest focal depths, of about 700 km (430 miles), are approached.

The deeper-focus earthquakes commonly occur in patterns called Benioff zones that dip into the Earth, indicating the presence of a subducting slab. Dip angles of these slabs average about 45°, with some shallower and others nearly vertical. Benioff zones coincide with tectonically active island arcs such as Japan, Vanuatu, Tonga, and the Aleutians, and they are normally but not always associated with deep ocean trenches such as those along the South American Andes. Exceptions to this rule include Romania and the Hindu Kush mountain system. In most Benioff zones, intermediate- and deep-earthquake foci lie in a narrow layer, although recent precise hypocentral locations in Japan and elsewhere show two distinct parallel bands of foci 20 km apart.

Aftershocks, Foreshocks and Swarms

Usually, a major or even moderate earthquake of shallow focus is followed by many lesser-size earthquakes close to the original source region. This is to be expected if the fault rupture producing a major earthquake does not relieve all the accumulated strain energy at once. In fact, this dislocation is liable to cause an increase in the stress and strain at a number of places in the vicinity of the focal region, bringing crustal rocks at certain points close to the stress at which fracture occurs. In some cases an earthquake may be followed by 1,000 or more aftershocks a day.

Sometimes a large earthquake is followed by a similar one along the same fault source within an hour or perhaps a day. An extreme case of this is multiple earthquakes. In most instances, however, the first principal earthquake of a series is much more severe than the aftershocks. In general, the number of aftershocks per day decreases with time. The aftershock frequency is roughly inversely proportional to the time since the occurrence of the largest earthquake of the series.

Most major earthquakes occur without detectable warning, but some principal earthquakes are preceded by foreshocks. In another common pattern, large numbers of small earthquakes may occur in a region for months without a major earthquake. In the Matsushiro region of Japan, for instance, there occurred between August 1965 and August 1967 a series of hundreds of thousands of earthquakes, some sufficiently strong (up to Richter magnitude 5) to cause property damage but no casualties. The maximum frequency was 6,780 small earthquakes on April 17, 1966. Such series of earthquakes are called earthquake swarms. Earthquakes associated with volcanic activity often occur in swarms, though swarms also have been observed in many nonvolcanic regions.

Types of Earthquakes

An earthquake is a sudden vibration of the earth surface due to the rapid release of stored energy between tectonic plates. There are four major types of earthquakes, which are:

Tectonic Earthquake

The earth crust is made up of unevenly shaped slab of rocks called tectonic plates. The energy stored here causes the tectonic plates to move towards away or push against each other. With time the stored energy and the movement of the tectonic plates build up the enormous pressure within the region between two plates.

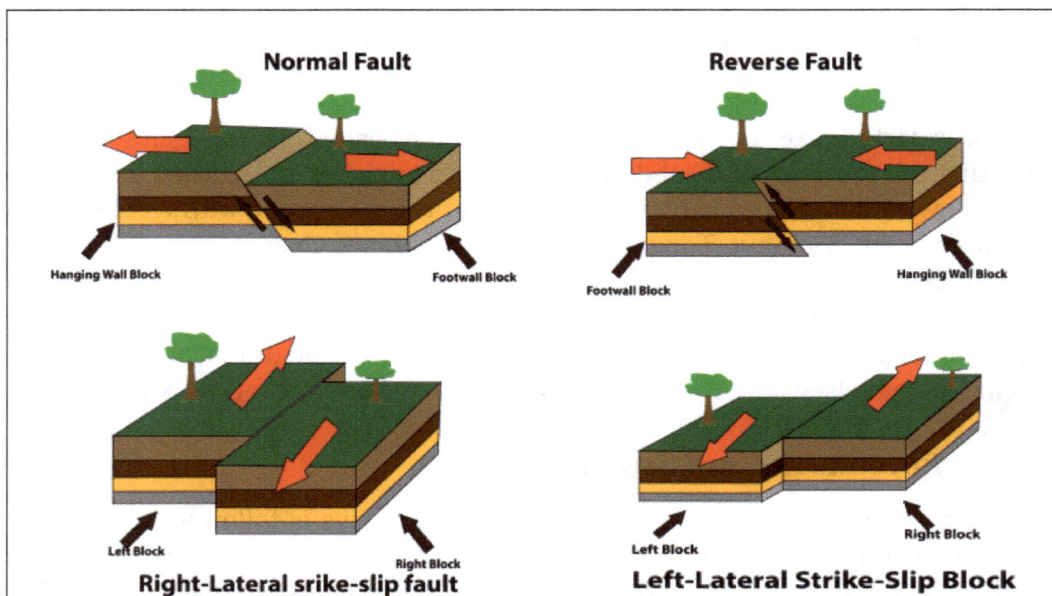

That intense pressure becomes the cause of the fault line(Like San Andrea Fault line), and plates move over against or apart from each other.

The center point of this disturbance is known as the focus of an earthquake. From the focus, waves of energy traveled towards the surface, shaking the surface within. The point the surface directly above the focus is called the epicenter.

From the epicenter of an earthquake, the energy waves traveled in a different direction on the earth surface causes vigorous movement on the surface of the earth which is known as an earthquake.

A powerful earthquake can destroy an entire city.

Volcanic Earthquake

Earthquake-related to the volcanic activity are called a volcanic earthquake. The magnitude of these quakes is usually weak.

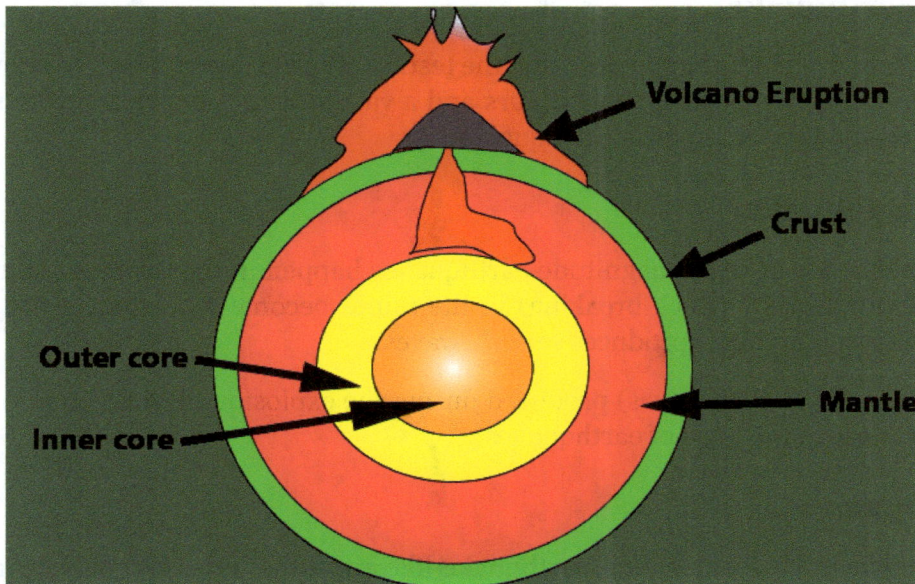

Volcanic earthquake.

Ever felt the largest volcanic earthquake was of the magnitude of 5.5.

- There are two types of volcanic earthquakes, which are:
  - ◦ Volcano-tectonic earthquake.
  - ◦ Long-period earthquake.

Volcano-tectonic Earthquake

The tremors happen due to injection or withdrawal of Magma between the stressed rocks is called a volcano-tectonic earthquake. The movements of the molten magma cause mostly volcanic earthquake directly underneath a volcano.

In these types of the earthquake, magma exerts the pressure on the tectonic plates until this magma breaks the rocks. During these cracks, tremors occur. These earthquakes are so weak that only can be measured by sensitive instruments.

After the breakdown of rocks, magma starts flowing towards the surface of the earth. After that, rocks have to fill up space where magma is no longer available. During filling the space, tremors happen of very low magnitude.

Long-period Earthquake

A long period earthquake occurs while the injection of magma into the surrounding rocks. These all happen due to the pressure changes among the layers of the earth. These types of volcanic activity indicate the eruption of the volcano in the near future.

Seismograph records these reading, and before the actual eruption of the volcano, we can easily take preventive measures. So that local people can become aware of the eruption of the volcano.

Explosive Earthquake

Explosive earthquake mostly happens during the testing of nuclear weapons. We know that during detonation of nuclear weapon big blast occurs and a vast amount of energy releases. Sometimes these blasts become the cause of the earthquakes.

Collapse Earthquake

These earthquakes are of weak magnitude earthquakes happen in the caverns and mines. Sometimes, underground blasts (Rock breaking) in the mines become the cause of the collapsing of mines and collapsing of mines produces seismic waves.

The P- and S-waves (Seismic waves) produced during the explosion of rocks on or under the surface of the earth cause this type of earthquakes.

Slow Earthquake

A slow earthquake is a discontinuous, earthquake-like event that releases energy over a period of hours to months, rather than the seconds to minutes characteristic of a typical earthquake. First detected using long term strain measurements, most slow earthquakes now appear to be accompanied by fluid flow and related tremor, which can be detected and approximately located using seismometer data filtered appropriately (typically in the 1–5 Hz band). That is, they are quiet compared to a regular earthquake, but not "silent" as described in the past.

Slow earthquakes should not be confused with tsunami earthquakes, in which relatively slow rupture velocity produces tsunami out of proportion to the triggering earthquake. In a tsunami earthquake, the rupture propagates along the fault more slowly than usual, but the energy release occurs on a similar timescale to other earthquakes.

Causes

Earthquakes occur as a consequence of gradual stress increases in a region, and once it reaches the maximum stress that the rocks can withstand a rupture generates and the resulting earthquake motion is related to a drop in the shear stress of the system. Earthquakes generate seismic waves when the rupture in the system occurs, the seismic waves consist of different types of waves that are capable of moving through the Earth like ripples over water. The causes that lead to slow earthquakes have only been theoretically investigated, by the formation of longitudinal shear cracks that were analysed using mathematical models. The different distributions of initial stress, sliding frictional stress, and specific fracture energy are all taken into account. If the initial stress minus the sliding frictional stress (with respect to the initial crack) is low, and the specific fracture energy or the strength of the crustal material (relative to the amount of stress) is high then slow earthquakes will occur regularly. In other words, slow earthquakes are caused by a variety of stick-slip and creep processes intermediated between asperity-controlled brittle and ductile fracture. Asperities are tiny bumps and protrusions along the faces of fractures. They are best documented from intermediate crustal levels of certain subduction zones (especially those that dip shallowly — SW Japan, Cascadia, Chile), but appear to occur on other types of faults as well, notably strike-slip plate boundaries such as the San Andreas fault and "mega-landslide" normal faults on the flanks of volcanos.

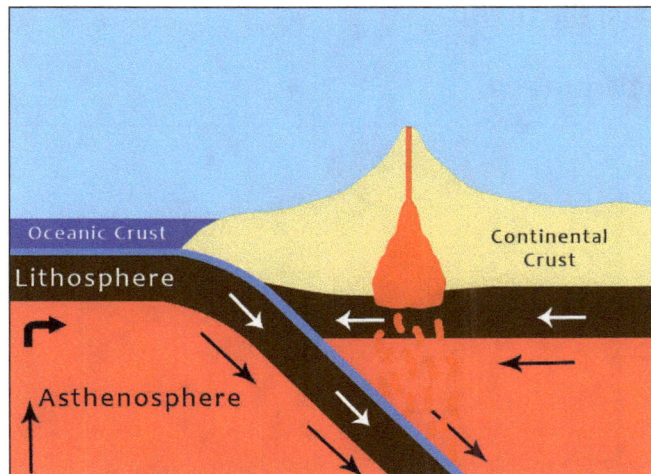

Common Cross Section of a Subduction Zone.

Locations

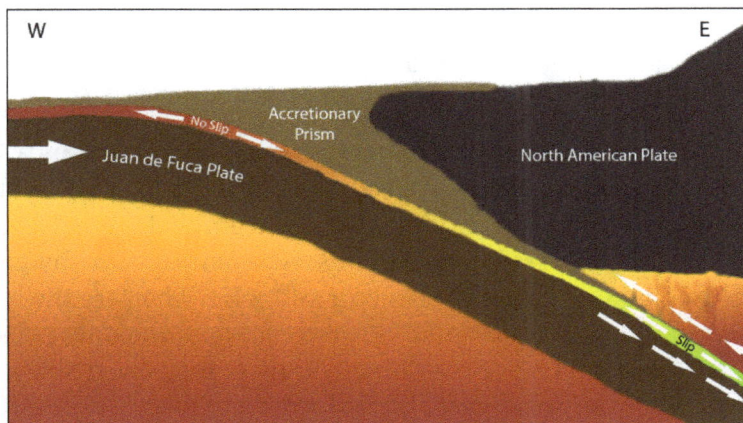

Cascadia Subduction Cross Section.

Faulting takes place all over Earth; faults can include convergent, divergent, and transform faults, and normally occur on plate margins. As of 2013 some of the locations that have been recently studied for slow earthquakes include: Cascadia, California, Japan, New Zealand, Mexico, and Alaska. The locations of slow earthquakes can provide new insights into the behavior of normal or fast earthquakes. By observing the location of tremors associated with slow-slip and slow earthquakes, seismologists can determine the extension of the system and estimate future earthquakes in the area of study.

Types

Teruyuki Kato identifies various types of slow earthquake:

- Low frequency earthquakes (LFE).

- Very low frequency earthquakes (VLF).

- Slow slip events (SSE).

- Episodic tremor and slip (ETS).

Low Frequency Earthquakes

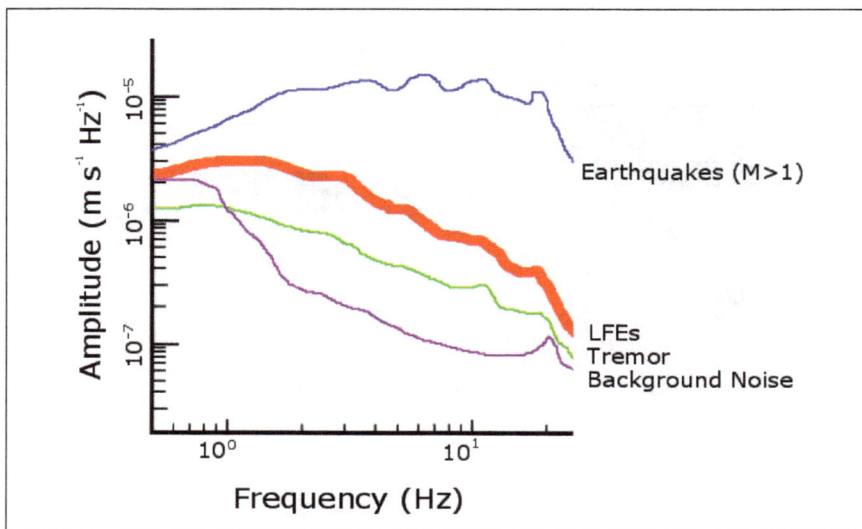

Plots of seismic events based on their average amplitudes and frequencies.
Low frequency earthquakes are peaked between 1 and 3 Hz.

Low frequency earthquakes (LFEs) are seismic events defined by waveforms with periods far greater than those of ordinary earthquakes and abundantly occur during slow earthquakes. LFEs can be volcanic, semi-volcanic, or tectonic in origin, but only tectonic LFEs or LFEs generated during slow earthquakes are described here. Tectonic LFEs are characterized by generally low magnitudes ($M<3$) and have frequencies peaked between 1 and 3 Hz. They are the largest constituent of non-volcanic tremor at subduction zones, and in some cases are the only constituent. In contrast to ordinary earthquakes, tectonic LFEs occur largely during long-lived slip events at subduction interfaces (up to several weeks in some cases) called slow slip events (SSEs). The mechanism responsible for their generation at subduction zones is thrust-sense slip along transitional segments of the plate interface. LFEs are highly sensitive seismic events which can likely be triggered by tidal forces as well as propagating waves from distant earthquakes. LFEs have hypocenters located down-dip from the seismogenic zone, the source region of megathrust earthquakes. During SSEs, LFE foci migrate along strike at the subduction interface in concert with the primary shear slip front.

The depth occurrence of low frequency earthquakes is in the range of approximately 20–45 kilometers depending on the subduction zone, and at shallower depths at strike-slip faults in California

At "warm" subduction zones like the west coast of North America, or sections in eastern Japan this depth corresponds to a transition or transient slip zone between the locked and stable slip intervals of the plate interface. The transition zone is located at depths approximately coincidental with the continental Mohorovicic discontinuity. At the Cascadia subduction zone, the distribution of LFEs form a surface roughly parallel to intercrustal seismic events, but displaced 5–10 kilometers down-dip, providing evidence that LFEs are generated at the plate interface.

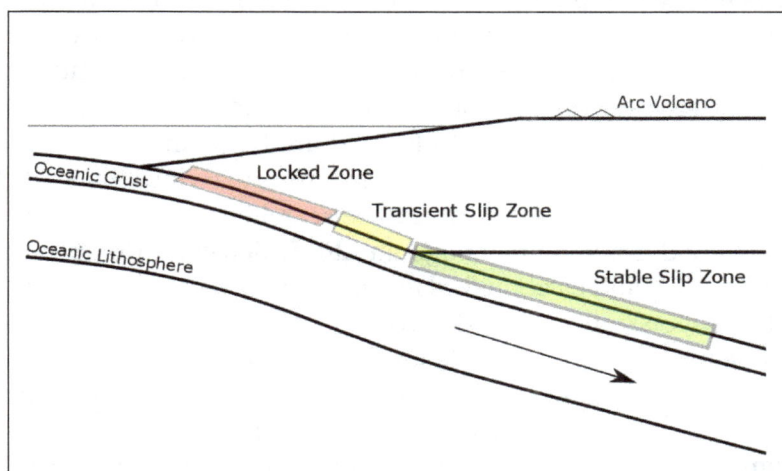

Subducting plate geometry and the kinematically defined interplate zones. The locked zone is the most shallow where the two plates are locked together, the transient slip zone is downdip of the locked zone and is the site of SSEs, and the stable slip zone is where the two plates are continuously slipping at their interface.

Low frequency earthquakes are an active area of research and may be important seismic indicators for higher magnitude earthquakes. Since slow slip events and their corresponding LFE signals have been recorded, none of them have been accompanied by a megathrust earthquake, however, SSEs act to increase the stress in the seismogenic zone by forcing the locked interval between the subducting and overriding plate to accommodate for down-dip movement. Some calculations find that the probability of a large earthquake occurring during a slow slip event are 30–100 times greater than background probabilities. Understanding the seismic hazard that LFEs might herald is among the primary reasons for their research. Additionally, LFEs are useful for the tomographic imaging of subduction zones because their distributions accurately map the deep plate contact near the Mohorovicic discontinuity.

Low frequency earthquakes were first classified in 1999 when the Japan Meteorological Agency (JMA) began differentiating LFE's seismic signature in their seismicity catalogue. The discovery and understanding of LFEs at subduction zones is due in part to the fact that the seismic signatures of these events were found away from volcanoes. Prior to their discovery, tremor events of this style were mainly associated with volcanism where the tremor is generated by partial coupling of flowing magmatic fluids. Japanese researchers first detected "low-frequency continuous tremor" near the top of the subducting Philippine Sea plate in 2002. After initially interpreting this seismic data as dehydration induced tremor, researchers in 2007 found that the data contained many LFE waveforms, or LFE swarms. Prior to 2007, tremor and LFEs were believed to be distinct events that often occurred together, but contemporarily LFEs are known to be the largest constituent forming tectonic tremor. LFEs and SSEs are frequently observed at subduction zones in western North America, Japan, Mexico, Costa Rica, New Zealand, as well as in shallow strike slip faults in California.

Detection

Low frequency earthquakes do not exhibit the same seismic character as regular earthquakes namely because they lack distinct, impulsive body waves. P-wave arrivals from LFEs have amplitudes so small that they are often difficult to detect, so when the JMA first distinguished the unique class of earthquake it was primarily by the detection of S-wave arrivals which were emergent. Because of this, detecting LFEs is nearly impossible using classical techniques. Despite their lack of important seismic identifiers, LFEs can be detected at low Signal-to-Noise-Ratio (SNR) thresholds using advanced seismic correlation methods. The most common method for identifying LFEs involves the correlation of the seismic record with a template constructed from confirmed LFE waveforms. Since LFEs are such subtle events and have amplitudes that are frequently drowned out by background noise, templates are built by stacking similar LFE waveforms to reduce the SNR. Noise is reduced to such an extent that a relatively clean waveform can be searched for in the seismic record, and when correlation coefficients are deemed high enough an LFE is detected. Determination of the slip orientation responsible for LFEs and earthquakes in general is done by the P-wave first-motion method. LFE P-waves, when successfully detected, have first motions indicative of compressional stress, indicating that thrust-sense slip is responsible for their generation. Extracting high quality P-wave data out of LFE waveforms can be quite difficult, however, and is furthermore important for accurate hypocentral depth determinations. The detection of high quality P-wave arrivals is a recent advent thanks to the deployment of highly sensitive seismic monitoring networks. The depth occurrence of LFEs are generally determined by P-wave arrivals but have also been determined by mapping LFE epicenters against subducting plate geometries. This method does not discriminate whether or not the observed LFE was triggered at the plate interface or within the down-going slab itself, so additional geophysical analysis is required to determine where exactly the focus is located. Both methods find that LFEs are indeed triggered at the plate contact.

Low Frequency Earthquakes in Cascadia

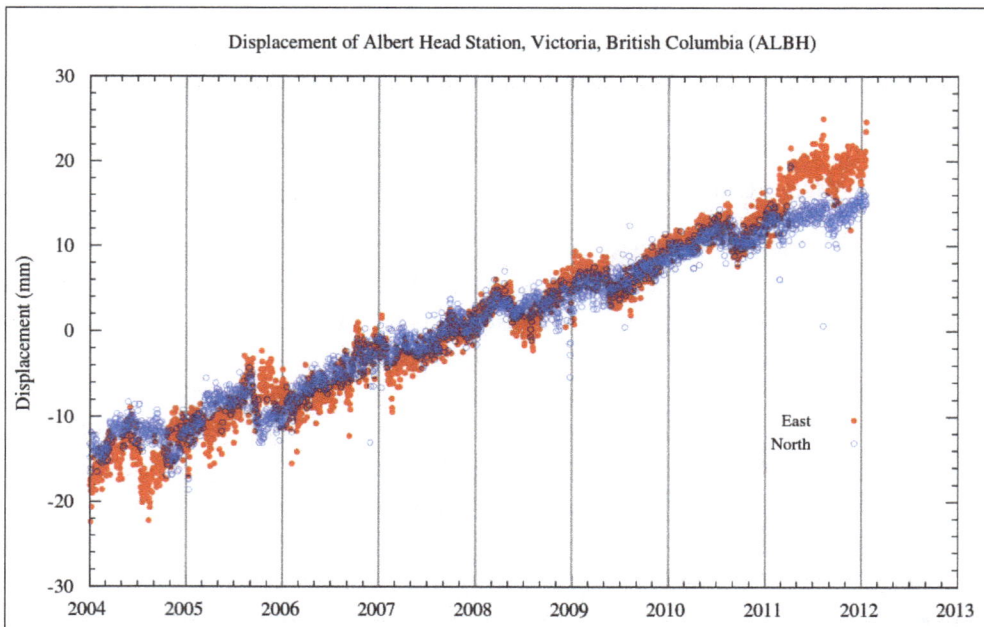

GPS data recording episodic slip events.

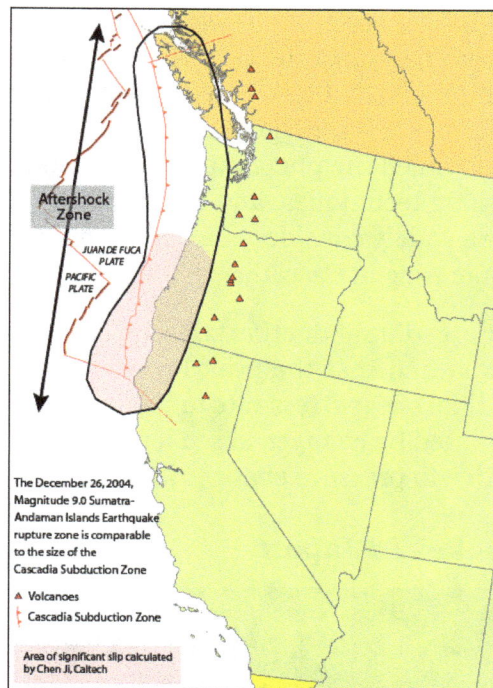

Cascadia subduction zone.

The Cascadia subduction zone spans from northern California to about halfway up Vancouver Island and is where the Juan de Fuca, Explorer, and Gorda plates are overridden by North America. In the Cascadia subduction zone, LFEs are predominantly observed at the plate interface down-dip of the seismogenic zone. In the southern section of the subduction zone from latitudes 40°N to 41.8°N low frequency earthquakes occur at depths between 28–47 kilometers, whereas farther north near Vancouver Island the range contracts to approximately 25–37 kilometers. This depth section of the subduction zone has been classified by some authors as the "transient slip" or "transition" zone due to its episodic slip behavior and is bounded up-dip and down-dip by the "locked zone" and "stable-slip zone", respectively. The transient slip section of the Cascadia is marked by high Vp/Vs ratios (P-wave velocity divided by S-wave velocity) and is designated as a Low Velocity Zone (LVZ). Furthermore, the LVZ has high Poisson's ratios as determined by teleseismic wave observations. These seismic properties defining the LVZ have been interpreted as an overpressured region of the down-going slab with high pore fluid pressures. The presence of water at the subduction interface and its relation to the generation of LFEs is not fully understood, but hydrolytic weakening of the rock contact is likely important.

Where megathrust earthquakes (M>8) have been repeatedly observed in the shallow sections (<25 km depth) of the Cascadia subduction zone, low frequency earthquakes have recently been discovered to occur at greater depths, down-dip of the seismogenic zone. The first indicator of low frequency earthquakes in Cascadia was discovered in 1999 when an aseismic event took place at the subduction interface wherein the overriding North American Plate slipped 2 centimeters southwest over a several week period as recorded by Global Positioning System (GPS) sites in British Columbia. This apparent slow slip event occurred over a 50-by-300-kilometer area and took approximately 35 days. Researchers estimated that the energy released in such an event would be equivalent to a magnitude 6–7 earthquake, yet no significant seismic signal was detected. The aseismic character of the event led observers to conclude that the slip was mediated by ductile deformation at depth. After further analysis of the GPS record, these reverse slip events were found to repeat at 13- to 16-month intervals, and last 2 to 4 weeks at any one GPS station. Soon after, geophysicists were able to extract the seismic signatures from these slow slip events and found that they were akin to tremor and classified the phenomenon as episodic tremor and slip (ETS). Upon the advent of improved processing techniques, and the discovery that LFEs form part of tremor, low frequency earthquakes were widely considered a commonplace occurrence at the plate interface down-dip of the seismogenic zone in Cascadia.

Low frequency tremors in the Cascadia subduction zone are strongly associated with tidal loading. A number of studies in Cascadia find that the peak low frequency earthquake signals alternate from being in phase with peak tidal shear stress rate to being in phase with peak tidal shear stress, suggesting that LFEs are modulated by changes in sea level. The shear slip events responsible for LFEs are therefore quite sensitive to pressure changes in the range of several kilo-pascals.

Low Frequency Earthquakes in Japan

Japan subduction setting.

The discovery of LFEs originates in Japan at the Nankai trough and is in part due to the nationwide collaboration of seismological research following the Kobe earthquake of 1995. Low frequency earthquakes in Japan were first observed in a subduction setting where the Philippine Sea plate subducts at the Nankai trough near Shikoku. The low-frequency continuous tremor researchers observed was initially interpreted to be a result of dehydration reactions in the subducting plate. The source of these tremors occurred at an average depth of around 30 kilometers, and they were distributed along the strike of the subduction interface over a length of 600 kilometers. Similar to Cascadia, these low frequency tremors occurred with slow slip events that had a recurrence interval of approximately 6 months. The later discovery of LFEs forming tremor confirmed the widespread existence of LFEs at Japanese subduction zones, and LFEs are widely observed and believed to occur as a result of SSEs.

The distribution of LFEs in Japan are centered around the subduction of the Philippine Sea plate and not the Pacific plate farther north. This is likely due to the difference in subduction geometries between the two plates. The Philippine Sea plate at the Nankai trough subducts at shallower over-all angles than does the Pacific plate at the Japan Trench, thereby making the Japan trench less suitable for SSEs and LFEs. LFEs in Japan have hypocenters located near the deepest extent of the transition zone, down-dip from the seismogenic zone. Estimates for the depth occurrence of the seismogenic zone near Tokai, Japan are 8–22 kilometers as determined by thermal methods. Fur-thermore, LFEs occur at a temperature range of 450–500 °C in Tokai, indicating that temperature may play an important role in the generation of LFEs in Japan.

Very Low Frequency Earthquakes

Very low frequency earthquakes (VLFs) can be considered a sub-category of low frequency earth-quakes that differ in terms of duration and period. VLFs have magnitudes of approximately 3-3.5, durations around 20 seconds, and are further enriched in low frequency energy (0.03–0.02 Hz). VLFs predominantly occur with LFEs, but the reverse is not true. There are two major subduction zone settings where VLFs have been detected, 1) within the offshore accretionary prism and 2) at the plate interface down-dip of the seismogenic zone. Since these two environments have consid-erably different depths, they have been termed shallow VLFs and deep VLFs, respectively. Like LFEs, very low frequency earthquakes migrate along-strike during ETS events. VLFs have been found at both the Cascadia subduction zone in western North America, as well as in Japan at the Nankai trough and Ryukyu trench. VLFs are produced by reverse fault mechanisms, similar to LFEs.

Slow Slip Events

Slow slip events (SSEs) are long lived shear slip events at subduction interfaces and the physical processes responsible for the generation of slow earthquakes. They are slow thrust-sense displace-ment episodes that can have durations up to several weeks, and are thus termed "slow". In many cases, the recurrence interval for slow slip events is remarkably periodic and accompanied by tec-tonic tremor, prompting seismologists to term episodic tremor and slip (ETS). In the Cascadia, the return period for SSEs is approximately 14.5 months, but varies along the margin of the sub-duction zone. In the Shikoku region in southwest Japan, the interval is shorter at approximately 6 months, as determined by crustal tilt changes. Some SSEs have durations in excess of several years, like the Tokai SSE that lasted from mid-2000 to 2003.

Slow slip event's locus of displacement propagate along the strike of subduction interfaces at ve-locities of 5–10 kilometers per day during slow earthquakes in the Cascadia, and this propagation is responsible for the similar migration of LFEs and tremor.

Episodic Tremor and Slip

Slow earthquakes can be episodic (relative of plate movement), and therefore somewhat predict-able, a phenomenon termed "episodic tremor and slip" or "ETS" in the literature. ETS events can last for weeks as opposed to "normal earthquakes" occur in a matter of seconds. Several slow-earth-quake events around the world appear to have triggered major, damaging seismic earthquakes in the shallower crust (e.g. 2001 Nisqually, 1995 Antofagasta). Conversely, major earthquakes trigger "post-seismic creep" in the deeper crust and mantle.

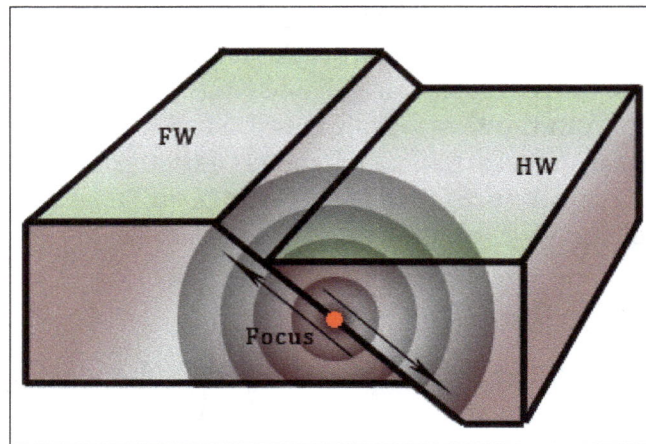

Earthquake FW-HW diagram.

Every five years a year-long quake of this type occurs beneath the New Zealand capital, Wellington. It was first measured in 2003, and has reappeared in 2008 and 2013. It lasts for around a year each time, releasing as much energy as a magnitude 7 quake.

Blind Thrust Earthquake

A blind thrust earthquake occurs along a thrust fault that does not show signs on the Earth's surface, hence the designation "blind". Such faults, being invisible at the surface, have not been mapped by standard surface geological mapping. Sometimes they are discovered as a by-product of oil exploration seismology; in other cases their existence is not suspected.

Although such earthquakes are not amongst the most energetic, they are sometimes the most destructive, as conditions combine to form an urban earthquake which greatly affects urban seismic risk.

A blind thrust earthquake is quite close, in meaning, to a buried rupture earthquake, if a buried rupture earthquake is not specifically about the fault, but signs the earthquake leaves, on the Earth's surface.

Blind Thrust Faults

Blind thrust faults generally exist near tectonic plate margins, in the broad disturbance zone. They form when a section of the Earth's crust is under high compressive stresses, due to plate margin collision, or the general geometry of how the plates are sliding past each other.

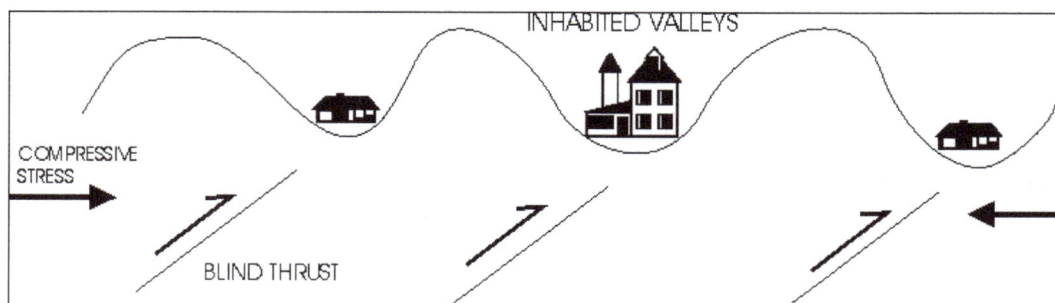

Diagram of blind-thrust faulting.

As shown in the diagram, a weak plate under compression generally forms thrusting sheets, or overlapping sliding sections. This can form a hill and valley landform, with the hills being the strong sections, and the valleys being the highly disturbed thrust faulted and folded sections. After a long period of erosion the visible landscape may be flattened, with material eroded from the hills filling up the valleys and hiding the underlying hill-and-valley geology. The valley rock is very weak and usually highly weathered, presenting deep, fertile soil; naturally, this is the area that becomes populated. Reflection seismology profiles show the disturbed rock that hides a blind thrust fault.

If the region is under active compression these faults are constantly rupturing, but any given valley might only experience a large earthquake every few hundred years. Although usually of magnitude 6 to 7 compared to the largest magnitude 9 earthquakes of recent times, such a temblor is especially destructive because the seismic waves are highly directed, and the soft basin soil of the valley can amplify the ground motions tenfold or more.

It is said that blind thrust earthquakes contribute more to urban seismic risk than the 'big ones' of magnitude 8 or more.

Examples of Occurrence

Some known Faults

- Los Angeles, California, USA, has many earthquakes and is well-studied. In addition to surface faults, a number of blind-thrust faults have been found under the basin and metropolitan area. A NASA study which combined satellite radar images and Global Positioning System (GPS) observations found that "tectonic squeezing across Los Angeles" "will likely produce earthquakes on either the blind Elysian Park or Puente Hills thrust fault systems".

- Bajo Segura Fault Zone, Spain.

- Fukaya Fault System, Japan (near Tokyo).

- Uemachi Fault System, Osaka Basin, Japan.

Specific Events

- 1987 Whittier Narrows earthquake.

- 1994 Northridge earthquake.

- 2010 Haiti earthquake.

- 2012 Visayas earthquake.

Megathrust Earthquake

Megathrust earthquakes occur at subduction zones at destructive convergent plate boundaries, where one tectonic plate is forced underneath another, caused by slip along the thrust fault that forms the contact between them. These interplate earthquakes are the planet's most powerful, with moment magnitudes (M_w) that can exceed 9.0. Since 1900, all earthquakes of magnitude 9.0 or greater have been megathrust earthquakes. No other type of known terrestrial source of tectonic activity has produced earthquakes of this scale.

During the rupture, one side of the fault is pushed upwards relative to the other, and it is this type of movement that is known as thrust. They are a type of dip-slip fault. A thrust fault is a reverse fault with a dip of 45° or less. Oblique-slip faults have significant components of different slip styles. The term *megathrust* does not have a widely accepted rigorous definition, but is used to refer to an extremely large thrust fault, typically formed at the plate interface along a subduction zone such as the Sunda megathrust.

Areas

Megathrust earthquakes are almost exclusive to tectonic subduction zones and are often associated with the Pacific and Indian Oceans. These subduction zones are not only responsible for megathrust earthquakes, but are also largely responsible for the volcanic activity associated with the Pacific Ring of Fire.

Since the earthquakes associated with these subduction zones deform the ocean floor, they often generate a significant series of tsunami waves. Subduction zone earthquakes are also known to produce intense shaking and ground movements for significant periods of time that can last for up to 5-6 minutes.

In the Indian Ocean region, the Sunda megathrust is located where the Indo-Australian Plate is subducting under the Eurasian Plate and extends 5,500 km (3300 mi) off the coasts of Myanmar, Sumatra, Java and Bali before terminating off the northwestern coast of Australia. This subduction zone was responsible for the 2004 Indian Ocean earthquake and tsunami.

In Japan, the Nankai megathrust under the Nankai Trough is responsible for Nankai megathrust earthquakes and associated tsunamis.

In the United States and Canada, the Juan de Fuca Plate is subducting under the North American Plate creating the Cascadia subduction zone which stretches from Middle Vancouver, British Columbia to Northern California. This subduction zone was responsible for the 1700 Cascadia earthquake.

A study reported in 2016 found that the largest megathrust quakes are associated with downgoing slabs with the shallowest dip, so-called flat slab subduction.

Table: Examples of megathrust earthquakes are listed in the following table.

Event	Estimated Magnitude (M_w)	Tectonic Plates Involved	Other Details/Notes
365 Crete earthquake	8.0+	African Plate subducting beneath the Aegean Sea Plate	• The quake generated a large tsunami in the eastern Mediterranean Sea and caused significant vertical displacement in the island of Crete. • Death toll in the "many thousands" according to Roman historian Ammianus Marcellinus.

869 Sanriku earthquake	8.6–9.0	Pacific Plate subducting beneath the Okhotsk Plate	• Slip length: 200 km over (125 mi over). • Slip width: 85 km over (53 mi over). • Led to a death toll of around 1000 people and completely destroyed the town of Tagajō.
1575 Valdivia earthquake	8.5	Nazca Plate subducting beneath the South American Plate	• Caused the dam at Laguna de Anigua (nowadays Riñihue Lake) to rupture and burst. Leading to the deaths of "many Native people" in Valdivia, Chile.
1700 Cascadia earthquake	8.7–9.2	Juan de Fuca Plate subducting beneath the North American Plate	• Slip length: 1000 km (625 mi). • Slip motion: 20 m (60 ft). • According to Native American oral traditions from the affected area, the quake and tsunami struck during a "winter evening" after many had gone to bed, wiping out several villages around Pachina Bay. A woman named Anacla Aq Sop was said to have been the only survivor of her entire tribe.
1868 Arica earthquake	8.5–9.0	Nazca Plate subducting beneath the South American Plate	• Slip length: 600 km (370 mi). • Resulted in near complete destruction of many cities and towns in the southern parts Peru. • The port city of Pisco was razed. • The tsunami that followed caused severe damage not only in Peru but also in Hawaii and caused the only recorded deaths by a tsunami in New Zealand. • Led to the deaths of over 25,000 people.
1906 Ecuador–Colombia earthquake	8.8		• The greatest damage from the tsunami occurred on the coast between Río Verde, Ecuador and Micay, Colombia. • Estimates of the number of deaths caused by the tsunami vary between 500 and 1,500.
2001 southern Peru earthquake	8.4	Nazca Plate subducting beneath the South American Plate	• Depth: 33 km. • Slip length: 300 km (190 mi). • Slip width: 120 km (75 mi).

Supershear Earthquake

A supershear earthquake is an earthquake in which the propagation of the rupture along the fault surface occurs at speeds in excess of the seismic shear wave (S-wave) velocity. This causes an effect analogous to a sonic boom.

Rupture Propagation Velocity

During seismic events along a fault surface the displacement initiates at the focus and then propagates outwards. Typically for large earthquakes the focus lies towards one end of the slip surface and much of the propagation is unidirectional (e.g. the 2008 Sichuan and 2004 Indian Ocean earthquakes). Theoretical studies have in the past suggested that the upper bound for propagation velocity is that of Rayleigh waves, approximately 0.92 of the shear wave velocity. However, evidence of propagation at velocities between S-wave and compressional wave (P-wave) values have been reported for several earthquakes in agreement with theoretical and laboratory studies that support the possibility of rupture propagation in this velocity range.

Occurrence

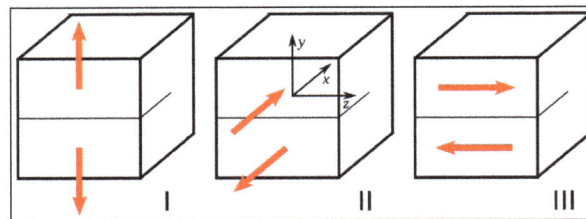

Mode-I, Mode-II, and Mode-III cracks.

Evidence of rupture propagation at velocities greater than S-wave velocities expected for the surrounding crust have been observed for several large earthquakes associated with strike-slip faults. During strike-slip, the main component of rupture propagation will be horizontal, in the direction of displacement, as a Mode II (in-plane) shear crack. This contrasts with a dip-slip rupture where the main direction of rupture propagation will be perpendicular to the displacement, like a Mode III (anti-plane) shear crack. Theoretical studies have shown that Mode III cracks are limited to the shear wave velocity but that Mode II cracks can propagate between the S and P-wave velocities and this may explain why supershear earthquakes have not been observed on dip-slip faults.

Initiation of Supershear Rupture

The rupture velocity range between those of Rayleigh waves and shear waves remains forbidden for a Mode II crack (a good approximation to a strike-slip rupture). This means that a rupture cannot accelerate from Rayleigh speed to shear wave speed. In the "Burridge–Andrews" mechanism, supershear rupture is initiated on a 'daughter' rupture in the zone of high shear stress developed at the propagating tip of the initial rupture. Because of this high stress zone, this daughter rupture is able start propagating at supershear speed before combining with the existing rupture. Experimental shear crack rupture, using plates of a photoelastic material, has produced a transition from sub-Rayleigh to supershear rupture by a mechanism that "qualitatively conforms to the well-known Burridge-Andrews mechanism".

Geological Effects

The high rates of strain expected near faults that are affected by supershear propagation are thought to generate what is described as pulverized rocks. The pulverization involves the development of many small microcracks at a scale smaller than the grain size of the rock, while preserving the earlier fabric, quite distinct from the normal brecciation and cataclasis found in most fault zones. Such rocks have been reported up to 400 m away from large strike-slip faults, such as the San Andreas Fault. The link between supershear and the occurrence of pulverized rocks is supported by laboratory experiments that show very high strain rates are necessary to cause such intense fracturing.

Examples:

Directly Observed

- 1999 Izmit earthquake, magnitude M_w 7.6 associated with strike-slip movement on the North Anatolian Fault Zone.

- 1999 Düzce earthquake, magnitude M_w 7.2 associated with strike-slip movement on the North Anatolian Fault Zone.

- 2001 Kunlun earthquake, magnitude M_w 7.8 associated with strike-slip movement on the Kunlun Fault.

- 2002 Denali earthquake, magnitude M_w 7.9 associated with strike-slip movement on the Denali Fault.

- 2010 Yushu earthquake, magnitude M_w 6.9 associated with strike-slip movement on the Yushu Fault.

- 2012 Indian Ocean earthquakes, magnitude M_w 8.6 associated with strike-slip on several fault segments - the first supershear event recognised in oceanic lithosphere.

- 2013 Craig earthquakes, magnitude M_w 7.6 associated with strike-slip on the Queen Charlotte Fault - the first supershear event recognised on an oceanic plate boundary.

- 2014 Aegean Sea earthquake, magnitude M_w 6.9, supershear was recognised during the second sub-event.

- 2015 Tajikistan earthquake, magnitude M_w 7.2, supershear slip on two segments, with normal slip at the restraining bend linking them.

- 2018 Sulawesi earthquake, magnitude M_w 7.5, associated with strike-slip movement on the Palu-Koro Fault.

Inferred

- 1906 San Francisco earthquake, magnitude M_w 7.8 associated with strike-slip movement on the San Andreas Fault.

- 1979 Imperial Valley earthquake, magnitude M_w 6.4 associated with slip on the Imperial Fault.

- 2013 Okhotsk Sea earthquake magnitude M_w 6.7 aftershock was an extremely deep (640 kilometers (400 miles)) supershear as well as unusually fast at "eight kilometers per second (five miles per second), nearly 50 percent faster than the shear wave velocity at that depth".

Submarine Earthquake

A submarine, undersea, or underwater earthquake is an earthquake that occurs underwater at the bottom of a body of water, especially an ocean. They are the leading cause of tsunamis. The magnitude can be measured scientifically by the use of the moment magnitude scale and the intensity can be assigned using the Mercalli intensity scale.

Understanding plate tectonics helps to explain the cause of submarine earthquakes. The Earth's surface or lithosphere comprises tectonic plates which average approximately 50 miles in thickness, and are continuously moving very slowly upon a bed of magma in the asthenosphere and inner mantle. The plates converge upon one another, and one subducts below the other, or, where there is only shear stress, move horizontally past each other. Little movements called fault creep are minor and not measurable. The plates meet with each other, and if rough spots cause the movement to stop at the edges, the motion of the plates continue. When the rough spots can no longer hold, the sudden release of the built-up motion releases, and the sudden movement under the sea floor causes a submarine earthquake. This area of slippage both horizontally and vertically is called the epicenter, and has the highest magnitude, and causes the greatest damage.

As with a continental earthquake the severity of the damage is not often caused by the earthquake at the rift zone, but rather by events which are triggered by the earthquake. Where a continental earthquake will cause damage and loss of life on land from fires, damaged structures, and flying objects; a submarine earthquake alters the seabed, resulting in a series of waves, and depending on the length and magnitude of the earthquake, tsunami, which bear down on coastal cities causing property damage and loss of life.

Submarine earthquakes can also damage submarine communications cables, leading to widespread disruption of the Internet and international telephone network in those areas. This is particularly common in Asia, where many submarine links cross submarine earthquake zones along Pacific Ring of Fire.

Tectonic Plate Boundaries

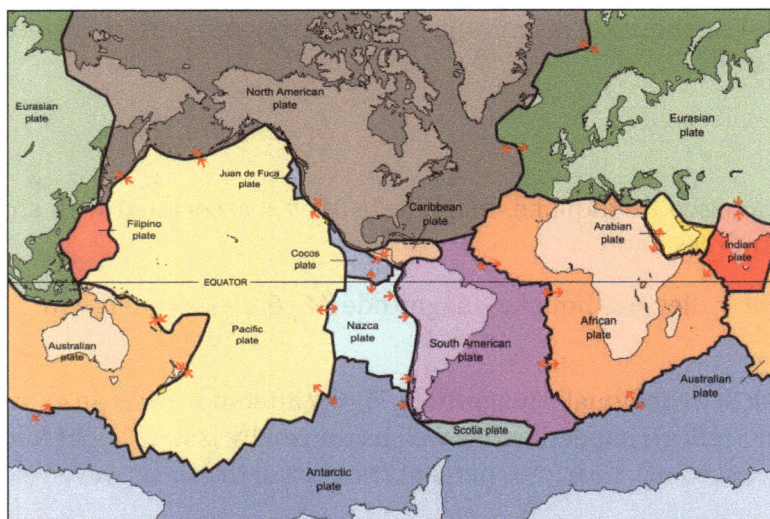

Tectonic plate boundaries, showing the directions of plate movements.

The different ways in which tectonic plates rub against each other under the ocean or sea floor to create submarine earthquakes. The type of friction created may be due to the characteristic of the geologic fault or the plate boundary as follows. Some of the main areas of large tsunami producing submarine earthquakes are the Pacific Ring of Fire and the Great Sumatran fault.

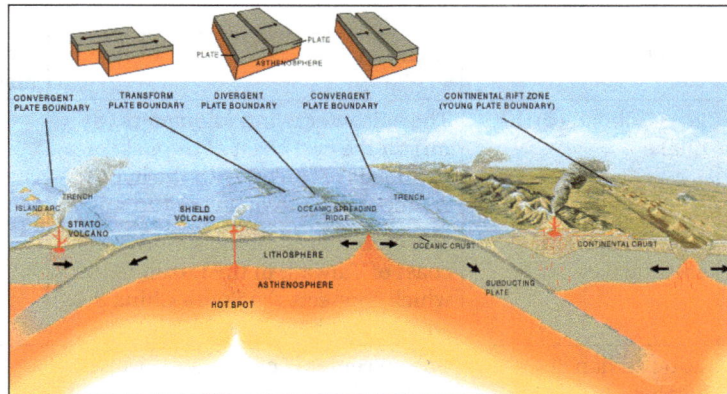

Different kinds of boundaries.

Convergent Plate Boundary

The older and denser plate moves below the lighter plate. The further down it moves, the hotter it becomes, until finally melting altogether at the asthenosphere and inner mantle and the crust is actually destroyed. The location where the two oceanic plates actually meet become deeper and deeper creating trenches with each successive action. There is an interplay of various densities of lithosphere rock, asthenosphere magma, cooling ocean water and plate movement for example the Pacific Ring of Fire. Therefore, the site of the sub oceanic trench will be a site of submarine earthquakes; for example the Mariana Trench, Puerto Rico Trench, and the volcanic arc along the Great Sumatran fault.

Transform Plate Boundary

A transform-fault boundary, or simply a transform boundary is where two plates will slide past each other, and the irregular pattern of their edges may catch on each other. The lithosphere is neither added to from the asthenosphere nor is it destroyed as in convergent plate action. For example, along the San Andreas fault strike-slip fault zone, the Pacific Tectonic Plate has been moving along at about 5 cm/yr in a northwesterly direction, whereas the North American Plate is moving south-easterly.

Divergent Plate Boundary

Rising convection currents occur where two plates are moving away from each other. In the gap, thus produced hot magma rises up, meets the cooler sea water, cools, and solidifies, attaching to either or both tectonic plate edges creating an oceanic spreading ridge. When the fissure again appears, again magma will rise up, and form new lithosphere crust. If the weakness between the two plates allows the heat and pressure of the asthenosphere to build over a large amount of time, a large quantity of magma will be released pushing up on the plate edges and the magma will solidify under the newly raised plate edges. If the fissure is able to come apart because of the two plates moving apart, in a sudden movement, an earthquake tremor may be felt for example at the Mid-Atlantic Ridge between North America and Africa.

List of Major Submarine Earthquakes

The following is a list of major submarine earthquakes since the 17th century:

Date	Event	Location	Estimated moment magnitude (M_w)
March 11, 2011	2011 Tōhoku earth-quake	The epicenter is 130 kilometers (81 mi) off the east coast of the Oshika Peninsula, Tōhoku, with the hypo-center at a depth of 32 km (20 mi).	9.1
December 26, 2006	2006 Hengchun earthquakes	The epicenter is off the southwest coast of Taiwan, in the Luzon Strait, which connects the South China Sea with the Philippine Sea.	7.1
December 26, 2004	2004 Indian Ocean earthquake	The epicenter is off the northwestern coast of Sumatra, Indonesia.	9.1
May 4, 1998		A part of the island of Yonaguni was destroyed by a submarine earth-quake.	
May 22, 1960	1960 Valdivia earth-quake	The epicenter is off the coast of South Central Chile.	9.5
December 20, 1946	1946 Nankaido earth-quake	The epicenter is off the southern coast of Kii Peninsula and Shikoku, Japan.	8.1
December 7, 1944	1944 Tōnankai earth-quake	The epicenter is about 20 km off the coast of the Shima Peninsula in Japan.	8.0
November 18, 1929	1929 Grand Banks earthquake	The epicenter is at Grand Banks, off the south coast of Newfoundland in the Atlantic Ocean.	7.2
June 15, 1896	1896 Sanriku earth-quake	The epicenter is off the Sanriku coast of northeastern Honshū, Japan.	8.5
April 4, 1771		The epicenter is near Yaeyama Is-lands in Okinawa, Japan.	7.4
January 26, 1700	1700 Cascadia earth-quake	The epicenter is offshore from Van-couver Island to northern California.	~9.0

Storm-caused Earthquakes

A 2019 study based on new higher-resolution data from the Transportable Array network of US-Array found that large ocean storms could create undersea earthquakes when they passed over certain areas of the ocean floor, including Georges Bank near Cape Cod and the Grand Banks of Newfoundland.

Deep-focus Earthquake

A deep-focus earthquake in seismology (also called a plutonic earthquake) is an earthquake with a hypocenter depth exceeding 300 km. They occur almost exclusively at convergent boundaries in association with subducted oceanic lithosphere. They occur along a dipping tabular zone beneath the subduction zone known as the Wadati-Benioff zone.

Seismic Characteristics

Deep-focus earthquakes give rise to minimal surface waves. Their focal depth causes the earthquakes to be less likely to produce seismic wave motion with energy concentrated at the surface. The path of deep-focus earthquake seismic waves from focus to recording station goes through the heterogeneous upper mantle and highly variable crust only once. Therefore, the body waves undergo less attenuation and reverberation than seismic waves from shallow earthquakes, resulting in sharp body wave peaks.

Focal Mechanisms

The pattern of energy radiation of an earthquake is represented by the moment tensor solution, which is graphically represented by beachball diagrams. An explosive or implosive mechanism produces an isotropic seismic source. Slip on a planar fault surface results in what is known as a double-couple source. Uniform outward motion in a single plane due to normal shortening gives rise is known as a compensated linear vector dipole source. Deep-focus earthquakes have been shown to contain a combination of these sources. The focal mechanisms of deep earthquakes depend on their positions in subducting tectonic plates. At depths greater than 400 km, down-dip compression dominates, while at depths of 250-300 km (also corresponding to a minimum in earthquake numbers vs. depth), the stress regime is more ambiguous but closer to down-dip tension.

Physical Process

Shallow-focus earthquakes are the result of the sudden release of strain energy built up over time in rock by brittle fracture and frictional slip over planar surfaces. However, the physical mechanism of deep focus earthquakes is poorly understood. Subducted lithosphere subject to the pressure and temperature regime at depths greater than 300 km should not exhibit brittle behavior, but should rather respond to stress by plastic deformation. Several physical mechanisms have been proposed for the nucleation and propagation of deep-focus earthquakes; however, the exact process remains an outstanding problem in the field of deep earth seismology.

The following four subsections outline proposals which could explain the physical mechanism allowing deep focus earthquakes to occur. With the exception of solid-solid phase transitions, the proposed theories for the focal mechanism of deep earthquakes hold equal footing in current scientific literature.

Solid-solid Phase Transitions

The earliest proposed mechanism for the generation of deep-focus earthquakes is an implosion due to a phase transition of material to a higher density, lower volume phase. The olivine-spinel phase transition is thought to occur at a depth of 410 km in the interior of the earth. This hypothesis proposes that metastable olivine in oceanic lithosphere subducted to depths greater than 410 km undergoes a sudden phase transition to spinel structure. The increase in density due to the reaction would cause an implosion giving rise to the earthquake. This mechanism has been largely discredited due to the lack of a significant isotropic signature in the moment tensor solution of deep-focus earthquakes.

Dehydration Embrittlement

Dehydration reactions of mineral phases with high weight percent water would increase the pore pressure in a subducted oceanic lithosphere slab. This effect reduces the effective normal stress in the slab and allow slip to occur on pre-existing fault planes at significantly greater depths that would normally be possible. Several workers suggest that this mechanism does not play a significant role in seismic activity beyond 350 km depth due to the fact that most dehydration reactions will have reached completion by a pressure corresponding to 150 to 300 km depth (5-10 GPa).

Transformational Faulting or Anticrack Faulting

Transformational faulting, also known as anticrack faulting, is the result of the phase transition of a mineral to a higher density phase occurring in response to shear stress in a fine-grained shear zone. The transformation occurs along the plane of maximal shear stress. Rapid shearing can then occur along these planes of weakness, giving rise to an earthquake in a mechanism similar to a shallow-focus earthquake. Metastable olivine subducted past the olivine-wadsleyite transition at 320-410 km depth (depending on temperature) is a potential candidate for such instabilities. Arguments against this hypothesis include the requirements that the faulting region should be very cold, and contain very little mineral-bound hydroxyl. Higher temperatures or higher hydroxyl contents preclude the metastable preservation of olivine to the depths of the deepest earthquakes.

Shear Instability or Thermal Runaway

A shear instability arises when heat is produced by plastic deformation faster than it can be conducted away. The result is thermal runaway, a positive feedback loop of heating, material weakening and strain-localisation within the shear zone. Continued weakening may result in partial melting along zones of maximal shear stress. Plastic shear instabilities leading to earthquakes have not been documented in nature, nor have they been observed in natural materials in the laboratory. Their relevance to deep earthquakes therefore lies in mathematical models which use simplified material properties and rheologies to simulate natural conditions.

Notable Deep-focus Earthquakes

The strongest deep-focus earthquake in seismic record was the magnitude 8.3 Okhotsk Sea earthquake that occurred at a depth of 609 km in 2013. The deepest earthquake ever recorded was a small 4.2 earthquake in Vanuatu at a depth of 735.8 km in 2004.

Remotely Triggered Earthquakes

Remotely triggered earthquakes are a result of the effects of large earthquakes at considerable distance, outside of the immediate aftershock zone. The further one gets from the initiating earthquake in both space and time, the more difficult it is to establish an association.

The physics involved in actually triggering an earthquake is complex. Most earthquake-generating zones are in a state of being close to failure. If such a zone were to be left completely alone, it would generate significant earthquakes spontaneously. Remote earthquakes, however, are in a position to disturb this critical state, either by shifting the stresses statically, or by dynamic change caused by passing seismic waves.

The first type of triggering may be due to static changes in the critical state. For example, after the magnitude 7.3 Landers earthquake struck California in 1992, it is said that "the earthquake map of California lit up like a Christmas tree". This event reinforced the idea of remotely triggered earthquakes, and pushed the hypothesis into the scientific mainstream. Following the very large 2004 Indian Ocean earthquake, it was established that remote earthquakes had been triggered as far away as Alaska.

There is scientific evidence for a "long reach", mainly in the form of discrete element modelling used in the mining industry. If rock is modeled as discrete elements in a critical state, a single disturbance can influence a wide area. A smaller-scale example is when a small excavation in a valley triggers a landslide and brings down a whole mountainside.

Doublet Earthquake

Doublet earthquakes – and more generally, multiplet earthquakes – were originally identified as multiple earthquakes with nearly identical waveforms originating from the same location. They are now characterized as single earthquakes having two (or more) main shocks of similar magnitude, sometimes occurring within tens of seconds, but sometimes separated by years. The similarity of magnitude – often within four-tenths of a unit of magnitude – distinguishes multiplet events from aftershocks, which start at about 1.2 magnitude less than the parent shock (Båth's law), and decrease in magnitude and frequency according to known laws.

Doublet/multiplet events also have nearly identical seismic waveforms, as they come from the same rupture zone and stress field, whereas aftershocks, being peripheral to the main rupture, typically reflect more diverse circumstances of origin. Multiplet events overlap in their focal fields (rupture zones), which can be up 100 kilometers across for magnitude 7.5 earthquakes. Doublets have been distinguished from *triggered earthquakes*, where the energy of the seismic waves triggers a distant earthquake with a different rupture zone, although it has been suggested such a distinction reflects "imprecise taxonomy" more than any physical reality.

Multiplet earthquakes are believed to result when *asperities*, such as large chunks of crust stuck in the rupturing fault, or irregularities or bends in the fault, temporarily impede the main rupture. Unlike a normal earthquake, where it is believed the earthquake releases enough of the tectonic stress driving it that it will take decades to centuries to accumulate enough stress to drive the next earthquake (per the elastic rebound theory), multiplet quakes have released only part of the pent-up stress when the rupture hits the asperity. This increases the stress across the asperity, which may fail within seconds, minutes, months, or even years. In the 1997 Harnai earthquake the initial M_w 7.0 shock was followed by an M_w 6.8 shock just 19 seconds later. The effect of such powerful shocks so close in time was to double the duration of ground shaking (bringing more structures to the point of collapse), and to double the area affected by the strongest shaking. When a subsequent, and possibly stronger, shock comes hours or days later it may suffice to collapse structures weakened by the previous shock, with serious consequences to rescue and recovery efforts.

Although there have been numerous earthquakes with two or even three primary shocks of such similar magnitude that picking one as the main shock can be somewhat arbitrary, it was not until the 1970s and 1980s that studies of seismograms showed that some of these were not simply unusually large foreshocks and aftershocks. Other studies have shown that about 20% of very large earthquakes (magnitude above 7.5) are doublets, and that, in some cases, 37 to 75 percent of earthquakes are multiplets. A theoretical study found about one earthquake in 15 (~7%) to be a doublet (using a narrow criterion of "doublet"), but also found that in the Solomon Islands six of 57 M ≥ 6.0 earthquakes were doublets, and 4 of 15 M ≥ 7.0 earthquakes, showing that approximately 10% and 25% of those quakes were doublets.

Doublet earthquakes pose a challenge to the characteristic earthquake model used for estimating seismic hazard. This model assumes that faults are segmented, limiting the extent of rupturing, and therefore the maximum size of an earthquake, to the length of the segment. Newer forecasts of seismic hazard, such as UCERF3, factor in a greater likelihood of multisegment ruptures, which changes the relative frequency of different sizes of earthquakes.

EARTHQUAKES AND FAULTS

When an earthquake occurs only a part of a fault is involved in the rupture. That area is usually outlined by the distribution of aftershocks in the sequence.

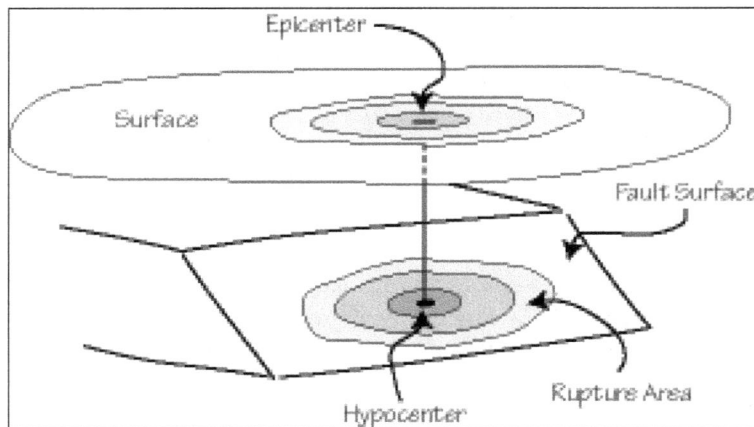

We call the "point" (or region) where an earthquake rupture initiates the hypocenter or focus. The point on Earth's surface directly above the hypocenter is called the epicenter. When we plot earthquake locations on a map, we usually center the symbol representing an event at the epicenter.

Generally, the area of the fault that ruptures increases with magnitude. Some estimates of rupture area are presented in the table below:

Date	Location	Length (km)	Depth (km)	(Mw)
04/18/06	San Francisco, CA	432	12	7.90
07/21/52	Kern County, CA	64	19	7.38
12/16/54	Fairview Peak, NV	50	15	7.17
12/16/54	Dixie Peak, NV	42	14	6.94
06/28/66	Parkfield, CA	35	10	6.25

02/09/71	San Fernando Valley, CA	17	14	6.64
10/28/83	Borah Peak, ID	33	20	6.93
10/18/89	Loma Prieta, CA	40	16	6.92
06/28/92	Landers, CA	62	12	7.34

Although the exact area associated with a given size earthquake varies from place to place and event to event, we can make predictions for "typical" earthquakes based on the available observations.

Magnitude	Fault Dimensions (Length x Depth, in km)
4.0	1.2 x 1.2
5.0	3.3 x 3.3
6.0	10 x 10
6.5	16 x 16, 25 x 10
7.0	40 x 20, 50 x 15
7.5	140 x 15, 100 x 20, 72 x 30, 50 x 40, 45 x 45
8.0	300 x 20, 200 x 30, 150 x 40, 125 x 50

These numbers should give you a rough idea of the size of structure that we are talking about when we discuss earthquakes.

Fault Structure

Although the number of observations of deep fault structure is small, the available exposed faults provide some information on the deep structure of a fault. A fault "zone" consists of several smaller regions defined by the style and amount of deformation within them.

The center of the fault is the most deformed and is where most of the offset or slip between the surrounding rock occurs. The region can be quite small, about as wide as a pencil is long, and it is identified by the finely ground rocks called cataclasite (we call the ground up material found closer to the surface, gouge). From all the slipping and grinding, the gouge is composed of very fine-grained material that resembles clay. Surrounding the central zone is a region several meters across that contains abundant fractures. Outside that region is another that contains distinguishable fractures, but much less dense than the preceding region. Last is the competent "host" rock that marks the end of the fault zone.

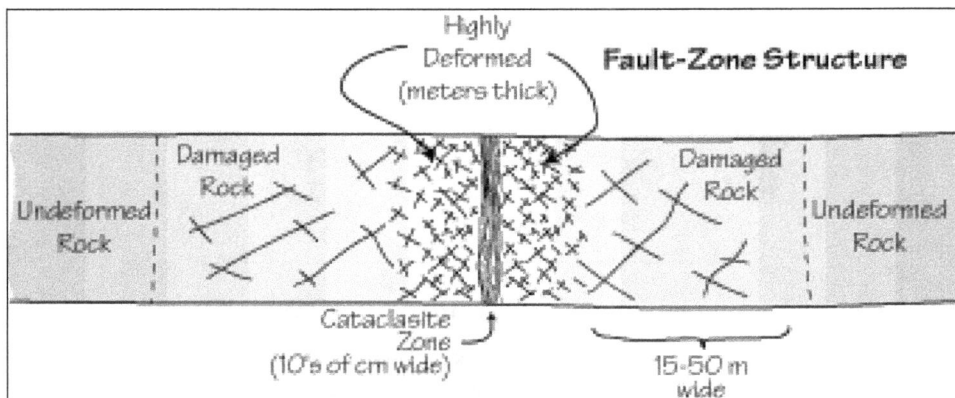

Structure of an exposed section of a vertical strike-slip fault zone.

Fault Classifications

Active, Inactive and Reactivated Faults

Active faults are structure along which we expect displacement to occur. By definition, since a shallow earthquake is a process that produces displacement across a fault, all shallow earthquakes occur on active faults.

Inactive faults are structures that we can identify, but which do not have earthquakes. As you can imagine, because of the complexity of earthquake activity, judging a fault to be inactive can be tricky, but often we can measure the last time substantial offset occurred across a fault. If a fault has been inactive for millions of years, it's certainly safe to call it inactive. However, some faults only have large earthquakes once in thousands of years, and we need to evaluate carefully their hazard potential.

Reactivated faults form when movement along formerly inactive faults can help to alleviate strain within the crust or upper mantle. Deformation in the New Madrid seismic zone in the central United States is a good example of fault reactivation. Structure formed about 500 Ma ago are responding to a new forces and relieving strain in the mid-continent.

Faulting Geometry

Faulting is a complex process and the variety of faults that exists is large. We will consider a simplified but general fault classification based on the geometry of faulting, which we describe by specifying three angular measurements: dip, strike, and slip.

Dip

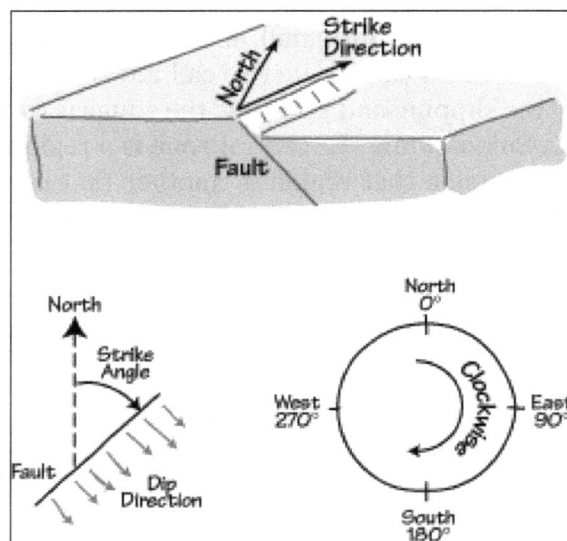

In Earth, faults take on a range of orientations from vertical to horizontal. Dip is the angle that describes the steepness of the fault surface. This angle is measured from Earth's surface, or a plane parallel to Earth's surface. The dip of a horizontal fault is zero (usually specified in degrees: 0°), and the dip of a vertical fault is 90°. We use some old mining terms to label the rock "blocks" above and below a fault. If you were tunneling through a fault, the material beneath the fault would be by your feet, the other material would be hanging above you head. The material resting on the fault is called the hanging wall, the material beneath the fault is called the foot wall.

Strike

The strike is an angle used to specify the orientation of the fault and measured clockwise from north. For example, a strike of 0° or 180° indicates a fault that is oriented in a north-south direction, 90° or 270° indicates east-west oriented structure. To remove the ambiguity, we always specify the strike such that when you "look" in the strike direction, the fault dips to you right. Of course if the fault is perfectly vertical you have to describe the situation as a special case. If a fault curves, the strike varies along the fault, but this is seldom causes a communication problem if you are careful to specify the location (such as latitude and longitude) of the measurement.

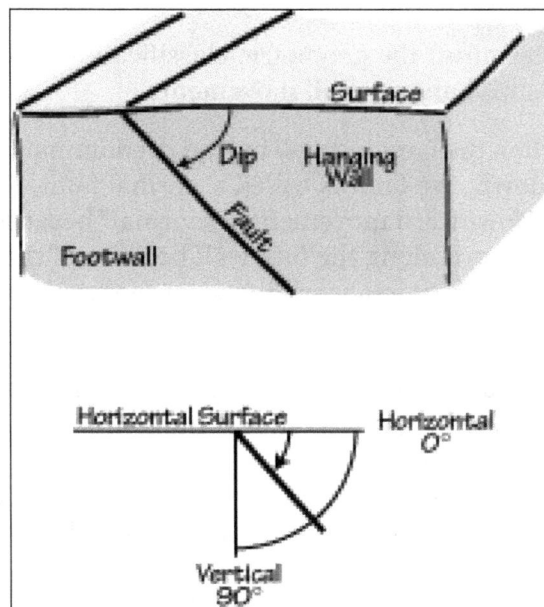

Slip

Dip and strike describe the orientation of the fault, we also have to describe the direction of motion across the fault. That is, which way did one side of the fault move with respect to the other. The parameter that describes this motion is called the slip. The slip has two components, a "magnitude" which tells us how far the rocks moved, and a direction (it's a vector). We usually specify the magnitude and direction separately.

The magnitude of slip is simply how far the two sides of the fault moved relative to one another; it's a distance usually a few centimeters for small earthquakes and meters for large events. The direction of slip is measured on the fault surface, and like the strike and dip, it is specified as an angle. Specifically the slip direction is the direction that the hanging wall moved relative to the footwall. If the hanging wall moves to the right, the slip direction is 0°; if it moves up, the slip angle is 90°, if it moves to the left, the slip angle is 180°, and if it moves down, the slip angle is 270° or -90°.

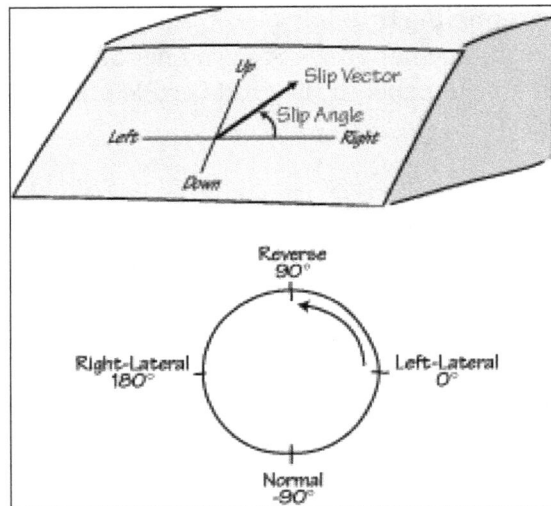

Hanging wall movement determines the geometric classification of faulting. We distinguish between "dip-slip" and "strike-slip" hanging-wall movements.

Dip-slip movement occurs when the hanging wall moved predominantly up or down relative to the footwall. If the motion was down, the fault is called a normal fault, if the movement was up, the fault is called a reverse fault. Downward movement is "normal" because we normally would expect the hanging wall to slide downward along the foot wall because of the pull of gravity. Moving the hanging wall up an inclined fault requires work to overcome friction on the fault and the downward pull of gravity.

When the hanging wall moves horizontally, it's a strike-slip earthquake. If the hanging wall moves to the left, the earthquake is called right-lateral, if it moves to the right, it's called a left-lateral fault. The way to keep these terms straight is to imagine that you are standing on one side of the fault and an earthquake occurs. If objects on the other side of the fault move to your left, it's a left-lateral fault, if they move to your right, it's a right-lateral fault.

When the hanging wall motion is neither dominantly vertical nor horizontal, the motion is called oblique-slip. Although oblique faulting isn't unusual, it is less common than the normal, reverse, and strike-slip movement.

Fault Styles

Faults and Forces

The style of faulting is an indicator of rock deformation and reflects the type of forces pushing or pulling on the region.

Near Earth's surface, the orientation of these forces are usually oriented such that one is vertical and the other two are horizontal. The precise direction of the horizontal forces varies from place to place as does the size of each force.

The style of faulting that is a reflection of the relative size of the different forces - in particular is the relative size of the vertical to the horizontal forces. There are three cases to consider, the vertical force can be the smallest, the largest, or the intermediate (neither smallest or largest). If the vertical force is the largest, we get normal faulting, if it is the smallest, we get reverse faulting. When the vertical force is the intermediate force, we get strike-slip faulting.

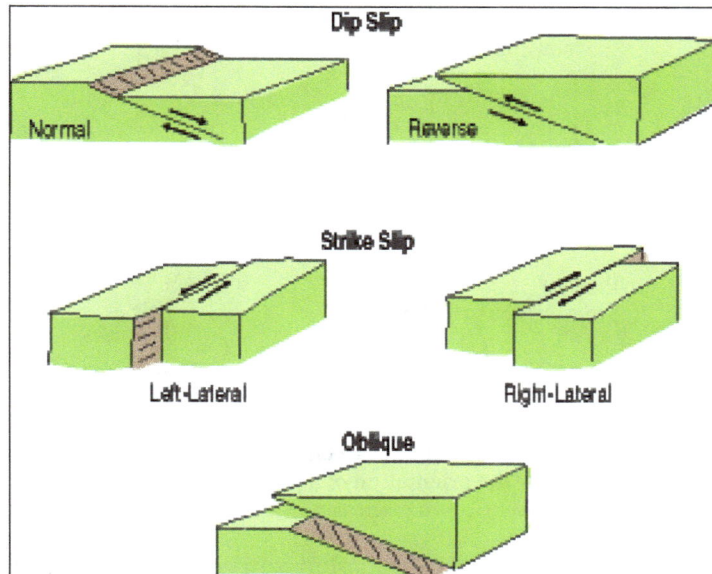

Normal faulting is indicative of a region that is stretching, and on the continents, normal faulting usually occurs in regions with relatively high elevation such as plateaus.

Reverse faulting reflects compressive forces squeezing a region and they are common in uplifting mountain ranges and along the coast of many regions bordering the Pacific Ocean. The largest earthquakes are generally low-angle (shallow dipping) reverse faults associated with "subduction" plate boundaries.

Strike-slip faulting indicates neither extension nor compression, but identifies regions where rocks are sliding past each other. The San Andreas fault system is a famous example of strike-slip deformation - part of coastal California is sliding to the northwest relative to the rest of North America - Los Angeles is slowly moving towards San Francisco.

As you might expect, the distribution of faulting styles is not random, but varies systematically across Earth and was one of the most important observations in constructing the plate tectonic model which explains so much of what we observe happening in the shallow part of Earth.

Fault Type:	Normal Faulting	Reverse Faulting	Transform Faulting
Deformation Style	Extension	Compression	Translation
Force Orientation	Vertical Force Is Largest	Vertical Force Is Smallest	Vertical Force Is Intermediate

Earthquake Focal Mechanisms

We use a specific set of symbols to identify faulting geometry on maps. The symbols are called earthquake focal mechanisms or sometimes "seismic beach balls". A focal mechanism is a graphical summary the strike, dip, and slip directions.

An earthquake focal mechanism is a projection of the intersection of the fault surface and an imaginary lower hemisphere (we'll use the lower hemisphere, but we could also use the upper hemisphere), surrounding the center of the rupture.

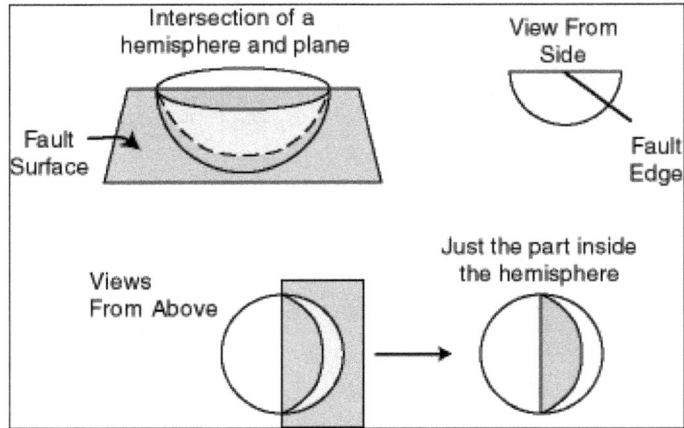

The intersection between the fault "plane" and the sphere is a curve. The focal mechanism shows the view of the hemisphere from directly above. We can show the orientation of a plane (i.e. the strike and dip) using just one curve, to include information on the slip, we use two planes and shade opposite quadrants of the hemisphere.

The price we pay for the ability to represent slip is that you cannot identify which of the two planes on the focal mechanism is the fault without additional information (such as the location and trend of aftershocks).

You should memorize the top three, which correspond to dip-slip reverse and normal faulting on a fault dipping 45°, and strike-slip faulting on a vertical fault. The lower two mechanisms correspond to a low-angle reverse earthquake (the dip is low) and the last example is an oblique event with components of both strike-slip and dip-slip movement. The strike of any plane can be read from a focal mechanism by identifying the intersection of the fault (shown as the boundary between shaded and unshaded regions) with the circle surrounding the mechanism (and using the dip-to-the-right rule).

Some example focal mechanisms are shown below.

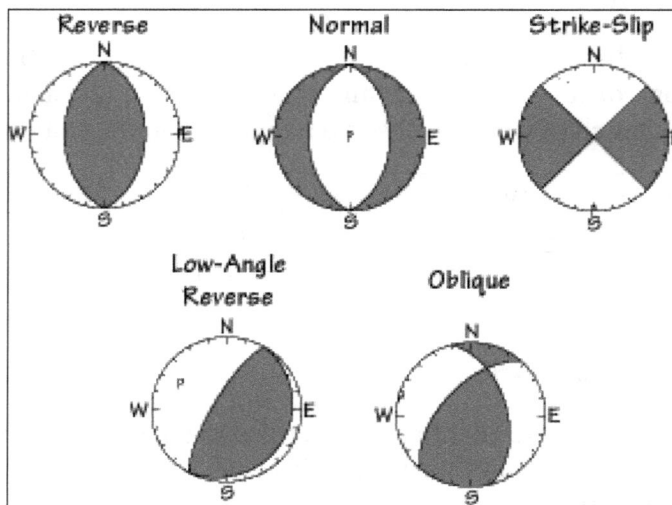

Stress and Strain

Stress is a force per unit area or a force that acts on a surface. Friction is a stress which resists motion and acts in all natural systems.

For earthquake studies, friction on faults and the orientation and relative magnitudes of the "regional" stresses that determine the style of faulting are of primary interest and importance.

Strain is a measure of material deformation such as the amount of compression when you squeeze or the amount of elongation when you stretch something.

In elastic deformation the amount of elongation is linearly proportional to the applied stress, and an elastic material returns to its original shape after the stress is relieved. Additionally, a strained, elastic material stores the energy used to deform it, and that energy is recoverable.

Elastic Rebound

Some regions repeatedly experience earthquakes and this suggests that perhaps earthquakes are part of a cycle. The effects of repeated earthquakes were first noted late in the nineteenth century by American geologist G. K. Gilbert. Gilbert observed a fresh fault scarp following the 1872 Owens Valley, California earthquake and correlated the scarp and uplift from a single earthquake with the uplift of the Sierra Nevada mountains. Decades later, following the 1906 San Francisco, California earthquake, H. F. Reid presented a similar hypothesis to explain better-documented movements along coastal California measured both before and after a large earthquake. Reid's model of the earthquake cycle has become known as the "Elastic Rebound Model".

The key to Reid's success was the availability of "before" and "after" observations for the earthquake which allowed him to see strain build in the crust before the event, and then see that strain release during the earthquake.

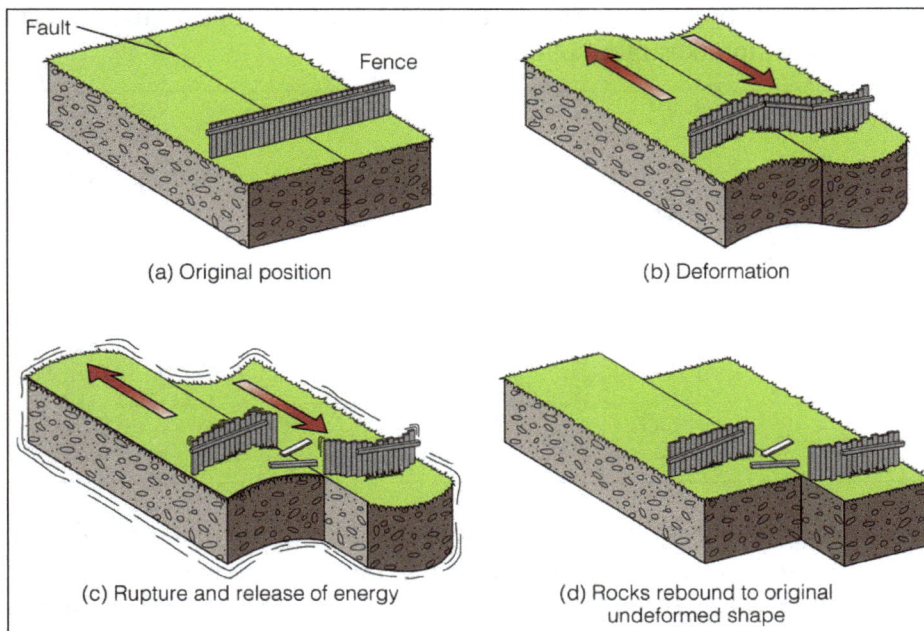

(a) Original position

(b) Deformation

(c) Rupture and release of energy

(d) Rocks rebound to original undeformed shape

In the diagram, we have two blocks of rock separated by a fault. As the two blocks move in opposite directions, friction acting on the fault resists movement and keeps the two sides from sliding. The rock strains as elastic energy is added, eventually, the strain loads the fault too much and overcomes the frictional "strength" of the fault. The rocks on either side of the fault jerk past each other in an earthquake. The earthquake releases the stored elastic strain energy as heat along the fault and as seismic vibrations.

For an ideal elastic-rebound fault, the stress on the fault periodically cycles between a minimum and maximum value and if the two blocks continue to move at a constant rate, the recurrence time (the time between earthquakes) is also uniform.

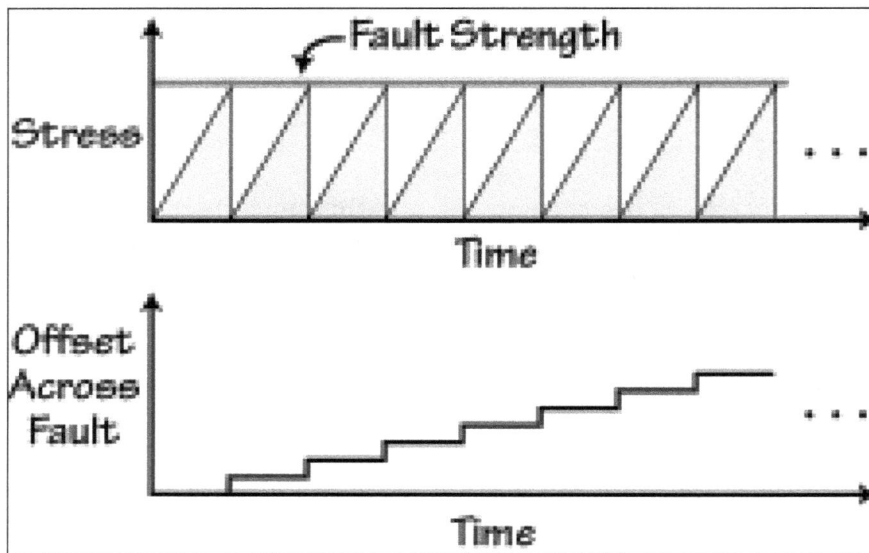

Unfortunately, actual faults are more complex, and the recurrence time is not periodic (which is one reason why earthquake prediction is so difficult). We have few observations of complete earthquake cycles because earthquakes take so long to recur.

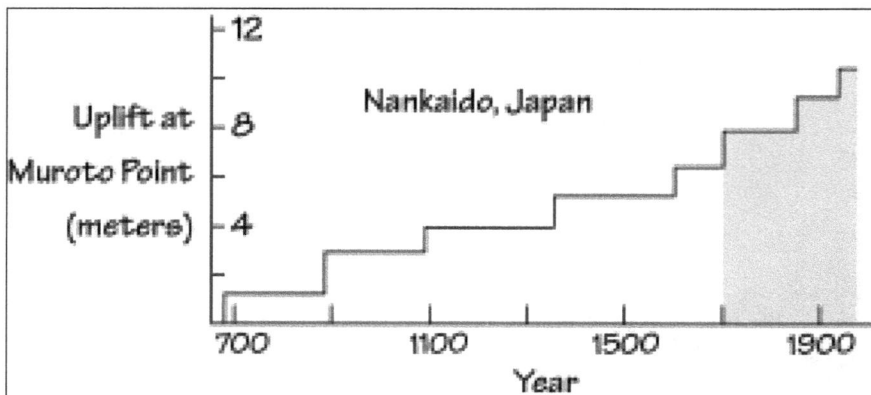

The figure above shows the observations from the Nankaido region of Japan (the gray region, the older values are estimated from earthquake histories), one of the few regions where observations on strain throughout several earthquake cycles exist. You can see that neither the time nor the slip is uniform from earthquake-to-earthquake.

EARTHQUAKE RUPTURE

An earthquake rupture is the extent of slip that occurs during an earthquake in the Earth's crust. Earthquakes occur for many reasons that include: landslides, movement of magma in a volcano, the formation of a new fault, or, most commonly of all, a slip on an existing fault.

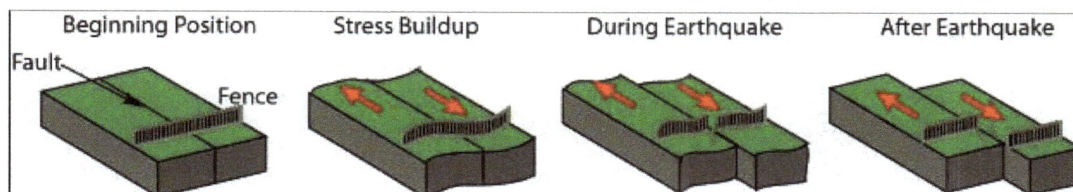

This cartoon shows what happens at the surface due to an earthquake rupture. Notice the progression of the strain that leads to the fault and amount of displacement.

Nucleation

A tectonic earthquake begins by an initial rupture at a point on the fault surface, a process known as nucleation. The scale of the nucleation zone is uncertain, with some evidence, such as the rupture dimensions of the smallest earthquakes, suggesting that it is smaller than 100 m while other evidence, such as a slow component revealed by low-frequency spectra of some earthquakes, suggest that it is larger. The possibility that the nucleation involves some sort of preparation process is supported by the observation that about 40% of earthquakes are preceded by foreshocks. However, some large earthquakes, such as the M8.6 1950 India - China earthquake, have no foreshocks and it remains unclear whether they just cause stress changes or are simply a result of increasing stresses in the region of the mainshock.

Once the rupture has initiated, it begins to propagate along the fault surface. The mechanics of this process are poorly understood, partly because it is difficult to recreate the high sliding velocities in a laboratory. Also the effects of strong ground motion make it very difficult to record information close to a nucleation zone.

Propagation

Following nucleation, the rupture propagates away from the hypocentre in all directions along the fault surface. The propagation will continue as long as there is sufficient stored strain energy to create new rupture surface. Although the rupture starts to propagate in all directions, it often becomes unidirectional, with most of the propagation in a mainly horizontal direction. Depending on the depth of the hypocentre, the size of the earthquake and whether the fault extends that far, the rupture may reach the ground surface, forming a surface rupture. The rupture will also propagate down the fault plane, in many cases reaching the base of the seismogenic layer, below which the deformation starts to become more ductile in nature.

Propagation may take place on a single fault, but in many cases the rupture starts on one fault before jumping to another, sometimes repeatedly. The 2002 Denali earthquake initiated on a thrust fault, the Sutsina Glacier Thrust, before jumping onto the Denali Fault for most of its propagation before finally jumping again onto the Totschunda Fault. The rupture of the 2016 Kaikoura earthquake was particularly complex, with surface rupture observed on at least 21 separate faults.

Termination

Some ruptures simply run out of sufficient stored energy, preventing further propagation. This may either be the result of stress relaxation due to an earlier earthquake on another part of the fault or because the next segment moves by aseismic creep, such that the stress never builds sufficiently to support rupture propagation. In other cases there is strong evidence for persistent barriers to propagation, giving an upper limit to earthquake magnitude.

Velocity

Most ruptures propagate at speeds in the range of 0.5–0.7 of the shear wave velocity, with only a minority of ruptures propagating significantly faster or slower than that.

The upper limit to normal propagation is the velocity of Rayleigh waves, 0.92 of the shear wave velocity, typically about 3.5 km per second. Observations from some earthquakes indicate that ruptures can propagate at speeds between the S-wave and P-wave velocity. These supershear earthquakes are all associated with strike-slip movement. The rupture cannot accelerate through the Rayleigh wave limit, so the accepted mechanism is that supershear rupture begins on a separate "daughter" rupture in the zone of high stress at the tip of the propagating main rupture. All observed examples show evidence of a transition to supershear at the point where the rupture jumps from one fault segment to another.

Slower than normal rupture propagation is associated with the presence of relatively mechanically weak material in the fault zone. This is particularly the case for some megathrust earthquakes, where the rupture velocity is about 1.0 km per second. These tsunami earthquakes are dangerous because most of the energy release happens at lower frequencies than normal earthquakes and they lack the peaks of seismic wave activity that would alert coastal populations to a possible tsunami risk. Typically the surface wave magnitude for such an event is much smaller than moment magnitude as the former does not capture the longer wavelength energy release. The 1896 Sanriku earthquake went almost unnoticed, but the associated tsunami killed more than 22,000 people.

Extremely slow ruptures take place on a time scale of hours to weeks, giving rise to slow earthquakes. These very slow ruptures occur deeper than the locked zone where normal earthquake ruptures occur on the same megathrusts.

EARTHQUAKE PREDICTION

Earthquake prediction is a branch of the science of seismology concerned with the specification of the time, location, and magnitude of future earthquakes within stated limits, and particularly the determination of parameters for the *next* strong earthquake to occur in a region. Earthquake prediction is sometimes distinguished from *earthquake forecasting*, which can be defined as the probabilistic assessment of *general* earthquake hazard, including the frequency and magnitude of damaging earthquakes in a given area over years or decades. Prediction can be further distinguished from earthquake warning systems, which upon detection of an earthquake, provide a real-time warning of seconds to neighboring regions that might be affected.

In the 1970s, scientists were optimistic that a practical method for predicting earthquakes would soon be found, but by the 1990s continuing failure led many to question whether it was even possible. Demonstrably successful predictions of large earthquakes have not occurred and the few claims of success are controversial. For example, the most famous claim of a successful prediction is that alleged for the 1975 Haicheng earthquake. A later study said that there was no valid short-term prediction. Extensive searches have reported many possible earthquake precursors, but, so far, such precursors have not been reliably identified across significant spatial and temporal scales. While part of the scientific community hold that, taking into account non-seismic precursors and given enough resources to study them extensively, prediction might be possible, most scientists are pessimistic and some maintain that earthquake prediction is inherently impossible.

Evaluating Earthquake Predictions

Predictions are deemed significant if they can be shown to be successful beyond random chance. Therefore, methods of statistical hypothesis testing are used to determine the probability that an earthquake such as is predicted would happen anyway (the null hypothesis). The predictions are then evaluated by testing whether they correlate with actual earthquakes better than the null hypothesis.

In many instances, however, the statistical nature of earthquake occurrence is not simply homogeneous. Clustering occurs in both space and time. In southern California about 6% of M≥3.0 earthquakes are "followed by an earthquake of larger magnitude within 5 days and 10 km". In central Italy 9.5% of M≥3.0 earthquakes are followed by a larger event within 48 hours and 30 km. While such statistics are not satisfactory for purposes of prediction (giving ten to twenty false alarms for each successful prediction) they will skew the results of any analysis that assumes that earthquakes occur randomly in time, for example, as realized from a Poisson process. It has been shown that a "naive" method based solely on clustering can successfully predict about 5% of earthquakes; "far better than 'chance'".

Option:	If Quake:	If No Quake:
Alarm	**Great losses, mitigated by** preparations (cost of alarm incidental).	**False alarm:** cost of alarm, panic and economic disruption. Multiple instances?
The Bar Lowering the bar...	reduces odds of losses	but increases the cost of false alarms.
No Alarm	**Great losses, worsened by** being caught off–guard.	**Normal:** no losses, no disruption, no cost of alarm.

The dilemma: To alarm? Or not to alarm?

As the purpose of short-term prediction is to enable emergency measures to reduce death and destruction, failure to give warning of a major earthquake, that does occur, or at least an adequate evaluation of the hazard, can result in legal liability, or even political purging. For example, it has been reported that members of the Chinese Academy of Sciences were purged for "having ignored scientific predictions of the disastrous Tangshan earthquake of summer 1976". Following the L'Aquila earthquake of 2009, seven scientists and technicians in Italy were convicted of manslaughter, but not so much for failing to *predict* the 2009 L'Aquila Earthquake (where some 300 people died) as for *giving undue assurance* to the populace – one victim called it "anaesthetizing" – that there would *not* be a serious earthquake, and therefore no need to take precautions. But warning of an earthquake that does not occur also incurs a cost: not only the cost of the emergency measures themselves, but of civil and economic disruption. False alarms, including alarms that are canceled, also undermine the credibility, and thereby the effectiveness, of future warnings. In 1999 it was reported that China was introducing "tough regulations intended to stamp out 'false' earthquake warnings, in order to prevent panic and mass evacuation of cities triggered by forecasts of major tremors". This was prompted by "more than 30 unofficial earthquake warnings in the past three years, none of which has been accurate". The acceptable trade-off between missed quakes and false alarms depends on the societal valuation of these outcomes. The rate of occurrence of both must be considered when evaluating any prediction method.

In a 1997 study of the cost-benefit ratio of earthquake prediction research in Greece, Stathis Stiros suggested that even a (hypothetical) excellent prediction method would be of questionable social utility, because "organized evacuation of urban centers is unlikely to be successfully accomplished", while "panic and other undesirable side-effects can also be anticipated." He found that earthquakes kill less than ten people per year in Greece (on average), and that most of those fatalities occurred in large buildings with identifiable structural issues. Therefore, Stiros stated that it would be much more cost-effective to focus efforts on identifying and upgrading unsafe buildings. Since the death toll on Greek highways is more than 2300 per year on average, he argued that more lives would also be saved if Greece's entire budget for earthquake prediction had been used for street and highway safety instead.

Prediction Methods

Earthquake prediction is an immature science — it has not yet led to a successful prediction of an earthquake from first physical principles. Researches into methods of prediction therefore focus on empirical analysis, with two general approaches: either identifying distinctive precursors to earthquakes, or identifying some kind of geophysical trend or pattern in seismicity that might precede a large earthquake. Precursor methods are pursued largely because of their potential utility for short-term earthquake prediction or forecasting, while trend methods are generally thought to be useful for forecasting, long term prediction (10 to 100 years time scale) or intermediate term prediction (1 to 10 years time scale).

Precursors

An earthquake precursor is an anomalous phenomenon that might give effective warning of an impending earthquake. Reports of these – though generally recognized as such only after the event – number in the thousands, some dating back to antiquity. There have been around 400 reports of possible precursors in scientific literature, of roughly twenty different types, running the gamut from aeronomy to zoology. None have been found to be reliable for the purposes of earthquake prediction.

In the early 1990, the IASPEI solicited nominations for a Preliminary List of Significant Precursors. Forty nominations were made, of which five were selected as possible significant precursors, with two of those based on a single observation each.

After a critical review of the scientific literature the *International Commission on Earthquake Forecasting for Civil Protection* (ICEF) concluded in 2011 there was "considerable room for methodological improvements in this type of research". In particular, many cases of reported precursors are contradictory, lack a measure of amplitude, or are generally unsuitable for a rigorous statistical evaluation. Published results are biased towards positive results, and so the rate of false negatives (earthquake but no precursory signal) is unclear.

Animal Behavior

For centuries there have been anecdotal accounts of anomalous animal behavior preceding and associated with earthquakes. In cases where animals display unusual behavior some tens of seconds prior to a quake, it has been suggested they are responding to the P-wave. These travel through the ground about twice as fast as the S-waves that cause most severe shaking. They predict not the earthquake itself — that has already happened — but only the imminent arrival of the more destructive S-waves.

It has also been suggested that unusual behavior hours or even days beforehand could be triggered by foreshock activity at magnitudes that most people do not notice. Another confounding factor of accounts of unusual phenomena is skewing due to "flashbulb memories": otherwise unremarkable details become more memorable and more significant when associated with an emotionally powerful event such as an earthquake. A study that attempted to control for these kinds of factors found an increase in unusual animal behavior (possibly triggered by foreshocks) in one case, but not in four other cases of seemingly similar earthquakes.

Dilatancy–diffusion

In the 1970s the dilatancy–diffusion hypothesis was highly regarded as providing a physical basis for various phenomena seen as possible earthquake precursors. It was based on "solid and repeatable evidence" from laboratory experiments that highly stressed crystalline rock experienced a change in volume, or *dilatancy*, which causes changes in other characteristics, such as seismic velocity and electrical resistivity, and even large-scale uplifts of topography. It was believed this happened in a 'preparatory phase' just prior to the earthquake, and that suitable monitoring could therefore warn of an impending quake.

Detection of variations in the relative velocities of the primary and secondary seismic waves – expressed as Vp/Vs – as they passed through a certain zone was the basis for predicting the 1973 Blue Mountain Lake (NY) and 1974 Riverside (CA) quake. Although these predictions were informal and even trivial, their apparent success was seen as confirmation of both dilatancy and the existence of a preparatory process, leading to what were subsequently called "wildly over-optimistic statements" that successful earthquake prediction "appears to be on the verge of practical reality".

However, many studies questioned these results, and the hypothesis eventually languished. Subsequent study showed it "failed for several reasons, largely associated with the validity of the assumptions on which it was based", including the assumption that laboratory results can be scaled up to the real world. Another factor was the bias of retrospective selection of criteria. Other studies have shown dilatancy to be so negligible that Main et al. 2012 concluded: "The concept of a large-scale preparation zone indicating the likely magnitude of a future event, remains as ethereal as the ether that went undetected in the Michelson–Morley experiment".

Changes in Vp/Vs

V_p is the symbol for the velocity of a seismic "P" (primary or pressure) wave passing through rock, while V_s is the symbol for the velocity of the "S" (secondary or shear) wave. Small-scale laboratory experiments have shown that the ratio of these two velocities – represented as V_p/V_s – changes when rock is near the point of fracturing. In the 1970s it was considered a likely breakthrough when Russian seismologists reported observing such changes in the region of a subsequent earthquake. This effect, as well as other possible precursors, has been attributed to dilatancy, where rock stressed to near its breaking point expands (dilates) slightly.

Study of this phenomenon near Blue Mountain Lake in New York State led to a successful albeit informal prediction in 1973, and it was credited for predicting the 1974 Riverside (CA) quake. However, additional successes have not followed, and it has been suggested that these predictions were a flukes. A V_p/V_s anomaly was the basis of a 1976 prediction of a M 5.5 to 6.5 earthquake near Los Angeles, which failed to occur. Other studies relying on quarry blasts (more precise, and repeatable) found no such variations, while an analysis of two earthquakes in California found that the variations reported were more likely caused by other factors, including retrospective selection of data. Geller noted that reports of significant velocity changes have ceased since about 1980.

Radon Emissions

Most rock contains small amounts of gases that can be isotopically distinguished from the normal atmospheric gases. There are reports of spikes in the concentrations of such gases prior to a major earthquake; this has been attributed to release due to pre-seismic stress or fracturing of the rock. One of these gases is radon, produced by radioactive decay of the trace amounts of uranium present in most rock.

Radon is useful as a potential earthquake predictor because it is radioactive and thus easily detected, and its short half-life (3.8 days) makes radon levels sensitive to short-term fluctuations. A 2009 review found 125 reports of changes in radon emissions prior to 86 earthquakes since 1966. But as the ICEF found in its review, the earthquakes with which these changes are supposedly linked were up to a thousand kilometers away, months later, and at all magnitudes. In some cases the anomalies were observed at a distant site, but not at closer sites. The ICEF found "no significant correlation". Another review concluded that in some cases changes in radon levels preceded an earthquake, but a correlation is not yet firmly established.

Electromagnetic Anomalies

Observations of electromagnetic disturbances and their attribution to the earthquake failure process go back as far as the Great Lisbon earthquake of 1755, but practically all such observations prior to the mid-1960s are invalid because the instruments used were sensitive to physical movement. Since then various anomalous electrical, electric-resistive, and magnetic phenomena have been attributed to precursory stress and strain changes that precede earthquakes, raising hopes for finding a reliable earthquake precursor. While a handful of researchers have gained much attention with either theories of how such phenomena might be generated, claims of having observed such phenomena prior to an earthquake, no such phenomena has been shown to be an actual precursor.

A 2011 review found the "most convincing" electromagnetic precursors to be ULF magnetic anomalies, such as the Corralitos event (discussed below) recorded before the 1989 Loma Prieta earthquake. However, it is now believed that observation was a system malfunction. Study of the closely monitored 2004 Parkfield earthquake found no evidence of precursory electromagnetic signals of any type; further study showed that earthquakes with magnitudes less than 5 do not produce significant transient signals. The International Commission on Earthquake Forecasting for Civil Protection (ICEF) considered the search for useful precursors to have been unsuccessful.

VAN Seismic Electric Signals

The most touted, and most criticized, claim of an electromagnetic precursor is the VAN method of physics professors Panayiotis Varotsos, Kessar Alexopoulos and Konstantine Nomicos (VAN) of the University of Athens. In a 1981 paper they claimed that by measuring geoelectric voltages – what they called "seismic electric signals" (SES) – they could predict earthquakes of magnitude larger than 2.8 within all of Greece up to seven hours beforehand.

In 1984 they claimed there was a "one-to-one correspondence" between SES and earthquakes – that is, that "*every sizable EQ is preceded by an SES* and inversely *every SES is always followed by an EQ* the magnitude and the epicenter of which can be reliably predicted" – the SES appearing between 6 and 115 hours before the earthquake. As proof of their method they claimed a series of successful predictions.

Although their report was "saluted by some as a major breakthrough" – one enthusiastic supporter (Uyeda) was reported as saying "VAN is the biggest invention since the time of Archimedes" – among seismologists it was greeted by a "wave of generalized skepticism". In 1996 a paper VAN submitted to the journal Geophysical Research Letters was given an unprecedented public peer-review by a broad group of reviewers, with the paper and reviews published in a special issue; the majority of reviewers found the methods of VAN to be flawed. Additional criticism was raised the same year in a public debate between some of the principals.

A primary criticism was that the method is geophysically implausible and scientifically unsound. Additional objections included the demonstrable falsity of the claimed one-to-one relationship of earthquakes and SES, the unlikelihood of a precursory process generating signals stronger than any observed from the actual earthquakes, and the very strong likelihood that the signals were man-made. Further work in Greece has tracked SES-like "anomalous transient electric signals" back to specific human sources, and found that such signals are not excluded by the criteria used by VAN to identify SES.

The validity of the VAN method, and therefore the predictive significance of SES, was based primarily on the empirical claim of demonstrated predictive success. Numerous weaknesses have been uncovered in the VAN methodology, and in 2011 the ICEF concluded that the prediction capability claimed by VAN could not be validated. Most seismologists consider VAN to have been "resoundingly debunked".

Corralitos Anomaly

Probably the most celebrated seismo-electromagnetic event ever, and one of the most frequently cited examples of a possible earthquake precursor, is the 1989 Corralitos anomaly. In the month prior to the 1989 Loma Prieta earthquake measurements of the earth's magnetic field at ultra-low frequencies by a magnetometer in Corralitos, California, just 7 km from the epicenter of the impending earthquake, started showing anomalous increases in amplitude. Just three hours before the quake the measurements soared to about thirty times greater than normal, with amplitudes tapering off after the quake. Such amplitudes had not been seen in two years of operation, nor in a similar instrument located 54 km away. To many people such apparent locality in time and space suggested an association with the earthquake.

Additional magnetometers were subsequently deployed across northern and southern California, but after ten years, and several large earthquakes, similar signals have not been observed. More recent studies have cast doubt on the connection, attributing the Corralitos signals to either unrelated magnetic disturbance or, even more simply, to sensor-system malfunction.

Freund Physics

In his investigations of crystalline physics, Friedemann Freund found that water molecules embedded in rock can dissociate into ions if the rock is under intense stress. The resulting charge carriers can generate battery currents under certain conditions. Freund suggested that perhaps these currents could be responsible for earthquake precursors such as electromagnetic radiation, earthquake lights and disturbances of the plasma in the ionosphere. The study of such currents and interactions is known as "Freund physics".

Most seismologists reject Freund's suggestion that stress-generated signals can be detected and put to use as precursors, for a number of reasons. First, it is believed that stress does not accumulate rapidly before a major earthquake, and thus there is no reason to expect large currents to be rapidly generated. Secondly, seismologists have extensively searched for statistically reliable electrical precursors, using sophisticated instrumentation, and have not identified any such precursors. And thirdly, water in the earth's crust would cause any generated currents to be absorbed before reaching the surface.

Trends

Instead of watching for anomalous phenomena that might be precursory signs of an impending earthquake, other approaches to predicting earthquakes look for trends or patterns that lead to an earthquake. As these trends may be complex and involve many variables, advanced statistical techniques are often needed to understand them, therefore these are sometimes called statistical methods. These approaches also tend to be more probabilistic, and to have larger time periods, and so merge into earthquake forecasting.

Elastic Rebound

Even the stiffest of rock is not perfectly rigid. Given a large force (such as between two immense tectonic plates moving past each other) the earth's crust will bend or deform. According to the elastic rebound theory of Reid (1910), eventually the deformation (strain) becomes great enough that something breaks, usually at an existing fault. Slippage along the break (an earthquake) allows the rock on each side to rebound to a less deformed state. In the process energy is released in various forms, including seismic waves. The cycle of tectonic force being accumulated in elastic deformation and released in a sudden rebound is then repeated. As the displacement from a single earthquake ranges from less than a meter to around 10 meters (for an M 8 quake), the demonstrated existence of large strike-slip displacements of hundreds of miles shows the existence of a long running earthquake cycle.

Characteristic Earthquakes

The most studied earthquake faults (such as the Nankai megathrust, the Wasatch fault, and the San Andreas fault) appear to have distinct segments. The *characteristic earthquake* model postulates that earthquakes are generally constrained within these segments. As the lengths and other properties of the segments are fixed, earthquakes that rupture the entire fault should have similar characteristics. These include the maximum magnitude (which is limited by the length of the rupture), and the amount of accumulated strain needed to rupture the fault segment. Since continuous plate motions cause the strain to accumulate steadily, seismic activity on a given segment should be dominated by earthquakes of similar characteristics that recur at somewhat regular intervals. For a given fault segment, identifying these characteristic earthquakes and timing their recurrence rate (or conversely return period) should therefore inform us about the next rupture; this is the approach generally used in forecasting seismic hazard. UCERF3 is a notable example of such a forecast, prepared for the state of California. Return periods are also used for forecasting other rare events, such as cyclones and floods, and assume that future frequency will be similar to observed frequency to date.

The idea of characteristic earthquakes was the basis of the Parkfield prediction: fairly similar earthquakes in 1857, 1881, 1901, 1922, 1934, and 1966 suggested a pattern of breaks every 21.9 years, with a standard deviation of ±3.1 years. Extrapolation from the 1966 event led to a prediction of an earthquake around 1988, or before 1993 at the latest (at the 95% confidence interval). The appeal of such a method is that the prediction is derived entirely from the *trend*, which supposedly accounts for the unknown and possibly unknowable earthquake physics and fault parameters. However, in the Parkfield case the predicted earthquake did not occur until 2004, a decade late.

This seriously undercuts the claim that earthquakes at Parkfield are quasi-periodic, and suggests the individual events differ sufficiently in other respects to question whether they have distinct characteristics in common.

The failure of the Parkfield prediction has raised doubt as to the validity of the characteristic earthquake model itself. Some studies have questioned the various assumptions, including the key one that earthquakes are constrained within segments, and suggested that the "characteristic earthquakes" may be an artifact of selection bias and the shortness of seismological records (relative to earthquake cycles). Other studies have considered whether other factors need to be considered, such as the age of the fault. Whether earthquake ruptures are more generally constrained within a segment (as is often seen), or break past segment boundaries (also seen), has a direct bearing on the degree of earthquake hazard: earthquakes are larger where multiple segments break, but in relieving more strain they will happen less often.

Seismic Gaps

At the contact where two tectonic plates slip past each other every section must eventually slip, as (in the long-term) none get left behind. But they do not all slip at the same time; different sections will be at different stages in the cycle of strain (deformation) accumulation and sudden rebound. In the seismic gap model the "next big quake" should be expected not in the segments where recent seismicity has relieved the strain, but in the intervening gaps where the unrelieved strain is the greatest. This model has an intuitive appeal; it is used in long-term forecasting, and was the basis of a series of circum-Pacific (Pacific Rim) forecasts in 1979 and 1989–1991.

However, some underlying assumptions about seismic gaps are now known to be incorrect. A close examination suggests that "there may be no information in seismic gaps about the time of occurrence or the magnitude of the next large event in the region"; statistical tests of the circum-Pacific forecasts shows that the seismic gap model "did not forecast large earthquakes well". Another study concluded that a long quiet period did not increase earthquake potential.

Seismicity Patterns

Various heuristically derived algorithms have been developed for predicting earthquakes. Probably the most widely known is the M8 family of algorithms (including the RTP method) developed under the leadership of Vladimir Keilis-Borok. M8 issues a "Time of Increased Probability" (TIP) alarm for a large earthquake of a specified magnitude upon observing certain patterns of smaller earthquakes. TIPs generally cover large areas (up to a thousand kilometers across) for up to five years. Such large parameters have made M8 controversial, as it is hard to determine whether any hits that happened were skillfully predicted, or only the result of chance.

M8 gained considerable attention when the 2003 San Simeon and Hokkaido earthquakes occurred within a TIP. In 1999, Keilis-Borok's group published a claim to have achieved statistically significant intermediate-term results using their M8 and MSc models, as far as world-wide large earthquakes are regarded. However, Geller are skeptical of prediction claims over any period shorter than 30 years. A widely publicized TIP for an M 6.4 quake in Southern California in 2004 was not fulfilled, nor two other lesser known TIPs. A deep study of the RTP method in 2008 found that out of some twenty alarms only two could be considered hits (and one of those had a 60% chance of happening anyway). It concluded that "RTP is not significantly different from a naïve method of guessing based on the historical rates of seismicity".

Accelerating moment release (AMR, "moment" being a measurement of seismic energy), also known as time-to-failure analysis, or accelerating seismic moment release (ASMR), is based on observations that foreshock activity prior to a major earthquake not only increased, but increased at an exponential rate. In other words, a plot of the cumulative number of foreshocks gets steeper just before the main shock.

Following formulation by Bowman into a testable hypothesis, and a number of positive reports, AMR seemed promising despite several problems. Known issues included not being detected for all locations and events, and the difficulty of projecting an accurate occurrence time when the tail end of the curve gets steep. But rigorous testing has shown that apparent AMR trends likely result from how data fitting is done, and failing to account for spatiotemporal clustering of earthquakes. The AMR trends are therefore statistically insignificant.

Difficulty or Impossibility

As the preceding examples show, the record of earthquake prediction has been disappointing. The optimism of the 1970s that routine prediction of earthquakes would be "soon", perhaps within ten years, was coming up disappointingly short by the 1990s, and many scientists began wondering why. By 1997 it was being positively stated that earthquakes can *not* be predicted, which led to a notable debate in 1999 on whether prediction of individual earthquakes is a realistic scientific goal.

Earthquake prediction may have failed only because it is "fiendishly difficult" and still beyond the current competency of science. Despite the confident announcement four decades ago that seismology was "on the verge" of making reliable predictions, there may yet be an underestimation of the difficulties. As early as 1978 it was reported that earthquake rupture might be complicated by "heterogeneous distribution of mechanical properties along the fault", and in 1986 that geometrical irregularities in the fault surface "appear to exert major controls on the starting and stopping of ruptures". Another study attributed significant differences in fault behavior to the maturity of the fault. These kinds of complexities are not reflected in current prediction methods.

Seismology may even yet lack an adequate grasp of its most central concept, elastic rebound theory. A simulation that explored assumptions regarding the distribution of slip found results "not in agreement with the classical view of the elastic rebound theory".

Earthquake prediction may be intrinsically impossible. It has been argued that the Earth is in a state of self-organized criticality "where any small earthquake has some probability of cascading into a large event". It has also been argued on decision-theoretic grounds that "prediction of major earthquakes is, in any practical sense, impossible".

That earthquake prediction might be intrinsically impossible has been strongly disputed. But the best disproof of impossibility – effective earthquake prediction – has yet to be demonstrated.

EARTHQUAKE SIMULATION

Earthquake simulation applies a real or simulated vibrational input to a structure that possesses the essential features of a real seismic event. Earthquake simulations are generally performed to study the effects of earthquakes on man-made engineered structures, or on natural features which may present a hazard during an earthquake.

Shake-table destructive testing of a model non-ductile 6-storey building.

Dynamic experiments on building and non-building structures may be physical – as with shake-table testing – or virtual (based on computer simulation). In all cases, to verify a structure's expected seismic performance, researchers prefer to deal with so called 'real time-histories' though the last cannot be 'real' for a hypothetical earthquake specified by either a building code or by some particular research requirements.

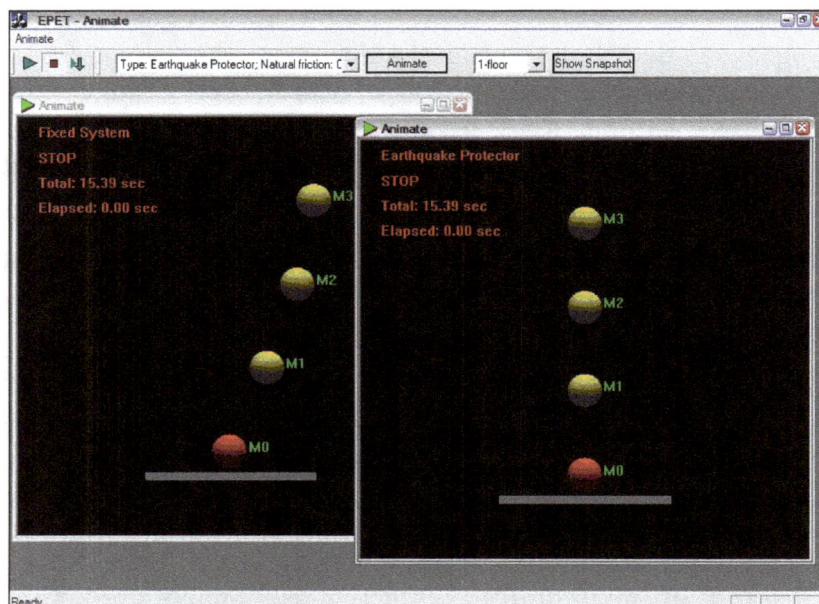

Print screen images of concurrent computer models animation.

Shake-table Testing

Studying a building's response to an earthquake is performed by putting a model of the structure on a shake-table that simulates the seismic loading. The earliest such experiments were performed more than a century ago.

Computational Approaches

Another way is to evaluate the earthquake performance analytically. The very first earthquake simulations were performed by statically applying some *horizontal inertia forces*, based on scaled peak ground accelerations, to a mathematical model of a building. With the further development of computational technologies, static approaches began to give way to dynamic ones.

Traditionally, numerical simulation and physical tests have been uncoupled and performed separately. So-called *hybrid testing* systems employ rapid, parallel analyses using both physical and computational tests.

EARTHQUAKE SWARM

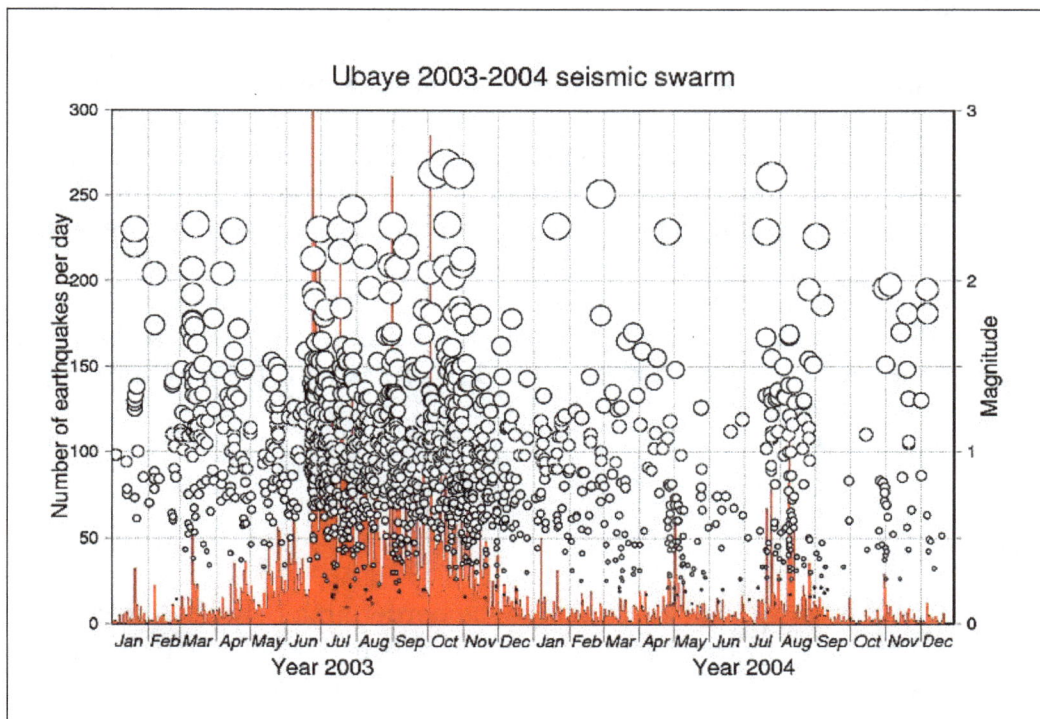

Chronology of the 2003-2004 Ubaye earthquake swarm.

An earthquake swarm is a sequence of seismic events occurring in a local area within a relatively short period of time. The length of time used to define the swarm itself varies, but may be of the order of days, months, or even years. Such an energy release is different from what happens commonly when a major earthquake (mainshock) is followed by a series of aftershocks: in earthquake swarms, no single earthquake in the sequence is obviously the mainshock. In particular, a cluster of aftershocks occurring after a mainshock is not a swarm.

The following examples were chosen for peculiarities of certain swarms (for instance: large number of events, complex interaction with larger shocks, long period of time, ultra-shallow focal depth), or because of their geographical region, some swarms occurring in otherwise aseismic regions. It is not intended to be a list of all the swarms happening worldwide.

Asia

India

- Since 11 November 2018, an earthquake swarm has been observed in the region of Dahanu, Maharashtra, an otherwise aseismic area. Ten to twenty quakes are felt daily, with magnitudes usually smaller than 3.5 (maximum magnitude 4.1 in February 2019). Even with this low-level of magnitude, two shocks proved destructive and even lethal, probably because their foci were very shallow.

Philippines

- An earthquake swarm occurred from early April 2017 to mid August 2017 in the Philippine province of Batangas. Four shocks in the 5.5–6.3 magnitude range (2017 Batangas earthquakes) caused damage in southern Luzon; they occurred at the beginning of the swarm: $M_s5.5$ (4 April), $M_s5.6$ and $M_s6.0$ (8 April), and $M_s6.3$ (11 April). The swarm origin of the 3 first major quakes seems established since they had practically the same epicentre; they occurred within the crust (7–28-km depth range). However, the strongest and lastest quake does not seem related to the swarm: its epicentre is 50 km away, and its focal depth is moreover very different (177 km, according to Phivolcs, the local seismic monitoring agency, a value which classifies this quake as an "intermediate-depth event"). This example shows how complex can be interaction between a swarm and an independent earthquake, even though this last one is very likely to have been triggered by the swarm activity.

Europe

Czechia/Germany

- The western Bohemia/Vogtland region is the border area between Czechia and Germany were earthquake swarms were first studied at the end of the 19th century. Swarm activity is recurrent there, sometimes with large maximum magnitudes, as for instance in 1908 (maximum magnitude 5.0), 1985–1986 (4.6), 2000 (3.2), or 2008 (3.8). This latter swarm occurred near Nový Kostel in October 2008 and lasted only 4 weeks, but up to 25,000 events were detected by WEBNET, the local monitoring network. The swarm is located on a steeply-dipping fault plane where an overall upward migration of activity was observed (first events at the bottom and last events at the top of the activated fault patch).

France

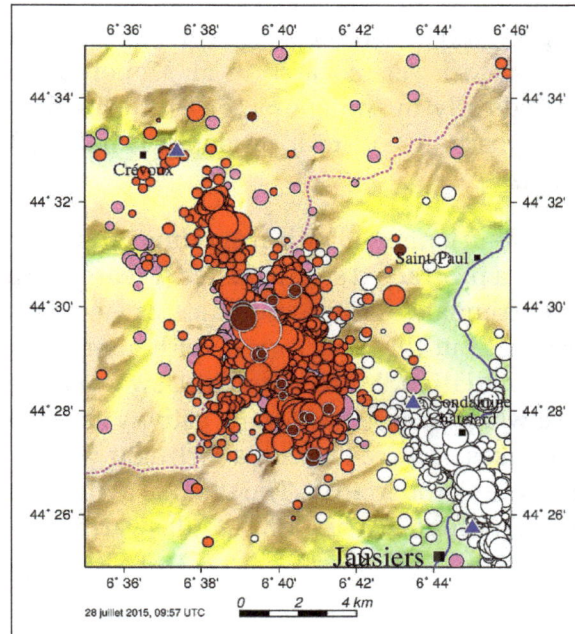

Ubaye earthquake swarms.

- In Alpes-de-Haute-Provence, the Ubaye Valley is the most active seismic zone in the French Alps. Earthquakes can follow there the classical scheme "mainshock + aftershocks" (for instance the 1959 M5.5 earthquake, which caused heavy damage and two casualities). But seismic energy is principally released by swarms. This is particularly the case in the upper valley, between Barcelonnette and the French-Italian border. At the beginning of the 21st century, La Condamine-Châtelard experienced an exceptional swarm activity in an area where usually only a few low-magnitude events occur every year. A first swarm developped in 2003–2004 when more than 16,000 events were detected by the local monitoring network, but with magnitudes keeping to low values (2.7). On a map, the 2003–2004 swarm is 8-km long. After a period of almost complete inactivity, it was followed by a second swarm (2012–2014), slightly offset by a few kilometres, and with a length of 11 km. This second swarm was initiated by an M4.3 earthquake in February 2012. Another M4.8 earthquake in April 2014 reactivated the swarm in 2014–2015. These two major shocks, which caused damage in the nearby localities, were of course followed by their own short sequence of aftershocks, but such a 4-year activity for moderate magnitude shocks clearly characterizes a swarm. Most foci were located in the 4–11-km depth range, within the crystalline basement. Focal mechanisms involve normal faulting, but also strike-slip faulting.

- In the lower Rhône Valley, the Tricastin has been known from the 18th century as the seat of earthquake swarms which sometimes caused damage, as in 1772–1773 and 1933–1936, and which were characterized by barrage-like detonations—at least so reported by the inhabitants. No seismic activity had been documented in the region since 1936, when a very weak swarm appeared for a few months in 2002–2003 (maximal magnitude 1.7). Had their foci not been sited just under a hamlet in the vicinity of Clansayes, and very close to the surface (200 m deep), these shocks would have gone unnoticed. In such a scenario of "ultra-shallow" seismicity, even earthquakes of very low magnitude (1, or 0, or even negative magnitude) can be felt as explosions or water-hammer noises, more than as vibrations. Most foci were located in an Upper-Cretaceous reef-limestone slab which bursts out periodically in the course of centuries for still unknown reasons for a few months or a few years. A 200-m focal depth is believed to be a worldwide record value for tectonic events.

- In the French Alps, the Maurienne Valley is from time to time prone to earthquake swarms. During the 19th century, a protracted swarm lasted 5 years and a half, from December 1838 to June 1844. Some earthquakes of the sequence caused damage in the region close to Saint-Jean-de-Maurienne, but this long swarm with many felt events made things particularly difficult for the population. More recently, a swarm appeared in October 2015 near Montgellafrey, in the lower part of the valley. Its activity kept low till 17 October 2017, when more than 300 earthquakes occurred within a fortnight, with a maximal magnitude of 3.7 being reached twice in late October 2017. The seismic activity lasted another full year, thus yielding a duration of more than 3 years for the full swarm.

Middle America

El Salvador

- In April 2017, the Salvadoran municipality of Antiguo Cuscatlán, a suburb of San Salvador, experienced a sequence of close to 500 earthquakes within 2 days, with magnitudes in the 1.5–5.1 range. There was one casuality and minor damage due to the strongest quake. Local experts *did not* identify any anomalous activity at nearby volcanoes.

Northern America

United States

- Between February and November 2008, Nevada experienced a swarm of 1,000 low-magnitude quakes generally referred to as the 2008 Reno earthquakes. The peak activity was in April 2008, when 3 quakes with magnitudes larger than 4 occurred within 2 days. The largest one registered 4.7 on the Richter scale and caused damage in the immediate area around the epicentre.

- The Yellowstone Caldera, a supervolcano in NW Wyoming, has experienced several strong earthquake swarms since the end of the 20th century. In 1985, more than 3,000 earthquakes were observed over a period of several months. More than 70 smaller swarms have been detected since. The United States Geological Survey states these swarms are likely caused by slips on pre-existing faults rather than by movements of magma or hydrothermal fluids. At the turn of the year 2008, more than 500 quakes were detected under the NW end of Yellowstone Lake over a seven-day span, with the largest registering a magnitude of 3.9. Another swarm started in January 2010, after the Haiti earthquake. With 1,620 small events in late January 2010, this swarm is the second-largest ever recorded in the Yellowstone Caldera. Interestingly, most of these swarms have "rapid-fire" characteristics.

They seemingly appear out of nowhere and can churn out tens or hundreds of small to moderate quakes within a very short time frame. Such swarms usually occur within the caldera boundary, as was especially the case in 2018.

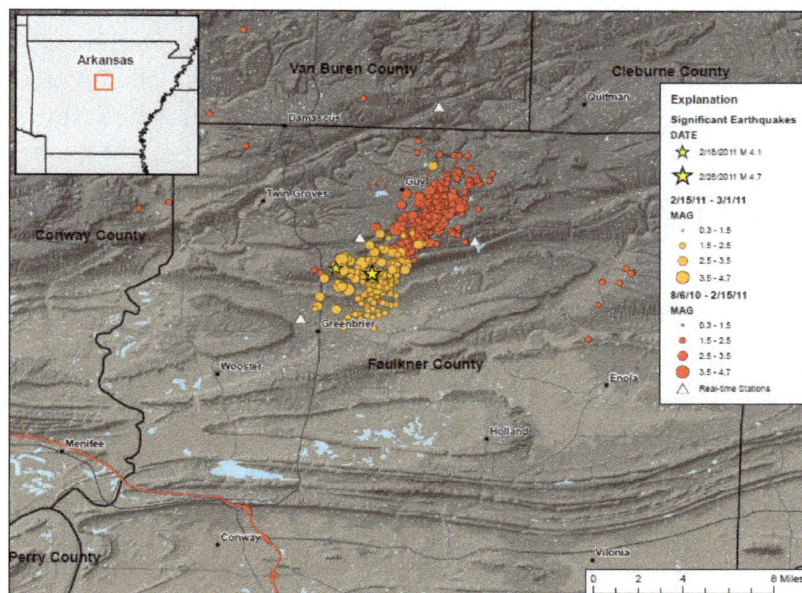

Guy-Greenbrier earthquake swarm: map of epicentres.

- The Guy-Greenbrier earthquake swarm occurred in central Arkansas beginning in August 2010. Epicentres show a linear distribution, with a clear overall shift in activity towards the southwest with time, and a magnitude of 4.7 was computed for the largest event. Analysis of the swarm has suggested a link with deep waste disposal drilling. It has led to a moratorium on such drilling.

- On 2 September 2017, an earthquake swarm appeared around Soda Springs, Idaho. Five quakes with magnitudes between 4.6 and 5.3 occurred within 9 days. Keeping the 2009 L'Aquila case in mind, and because Idaho had experienced an $M6.9$ earthquake in 1983, experts warned residents that a stronger quake could follow (an unlikely but still possible scenario for them).

Atlantic Ocean

- In El Hierro, the smallest and farthest south and west of the Canary Islands, hundreds of small earthquakes were recorded from July 2011 until October 2011 during the 2011–12 El Hierro eruption. The accumulated energy released by the swarm increased dramatically on 28 September. The swarm was due to the movement of magma beneath the island, and on 9 October a submarine volcanic eruption was detected.

Indian Ocean

- An intriguing earthquake swarm began east of Mayotte on 10 May 2018. The strongest quake ($M5.9$), the largest-magnitude event ever recorded in the Comoro zone, stroke on 15 May 2018. The swarm includes thousands of quakes, many of them felt by Maorais residents. Temporarily-installed ocean-bottom seismometers showed that the swarm active zone was sited 10 km east of Mayotte, deep into the oceanic lithosphere (in the 20–50-km depth range), a rather surprising result because the swarm was believed to be caused by the deflation of a magma reservoir located 45 km east of Mayotte, at a depth of 28 km. (Accordingly, an oceanographic campaign discovered in May 2019 a new submarine volcano, 800-m high and located 50 km east of Mayotte.) The swarm had been tapering off between August and November 2018 when the 11-November-2018 event occurred. This event had no detectable P nor S waves, but generated surface waves which could be observed world-wide by seismological observatories. Its origin is thought to be east of Mayotte. The swarm has continued to be active all through 2019.

Pacific Ocean

- In January and February 2013, the Santa Cruz Islands experienced a large earthquake swarm with many magnitude 5 and 6 earthquakes: More than 40 quakes with magnitude 4.5 or larger took place during the previous 7 days, including 7 events with magnitude larger than 6. The swarm degenerated into the M8.0 2013 Solomon Islands earthquake.

EARTHQUAKE ENGINEERING

Earthquake engineering is the science of the performance of buildings and structures when subjected to seismic loading. It also assists analysing the interaction between civil infrastructure and the ground, including the consequences of earthquakes on structures. One of the most important aims of earthquake engineering is the proper design and construction of buildings in accordance with building codes, so as to minimize damage due to earthquakes. It is the earthquake engineer who ensures proper design of buildings so they will resist damage due to earthquakes, but at the same time not be unnecessarily expensive.

Seismic Vibration Control Technologies

The purpose of these technologies is to minimize the seismic effects on buildings and other infrastructure by the use of seismic control devices. When seismic waves start penetrating the base of the buildings from the ground level, the flow density of their energy reduces due to reflections and other reasons. However, the remaining waves possess significant potential for damage when they reach the superstructure.

Vibration control devices assist in the reduction of the damaging effects, and enhance the seismic performance characteristics of the building. When the seismic waves penetrate a superstructure, these are dissipated by the use of dampers, or dispersed in a wide range of frequencies. Mass dampers are also employed to absorb the resonant wave frequencies of seismic waves, thus reducing the damaging effects. Seismic isolation techniques are sometimes used to partly suppress the flow of seismic energy into the superstructure by the insertion of pads into or beneath the load bearing elements in the base of the structure. Thus, the structure is protected from the damaging consequences of an earthquake by decoupling the structure from the shaking ground.

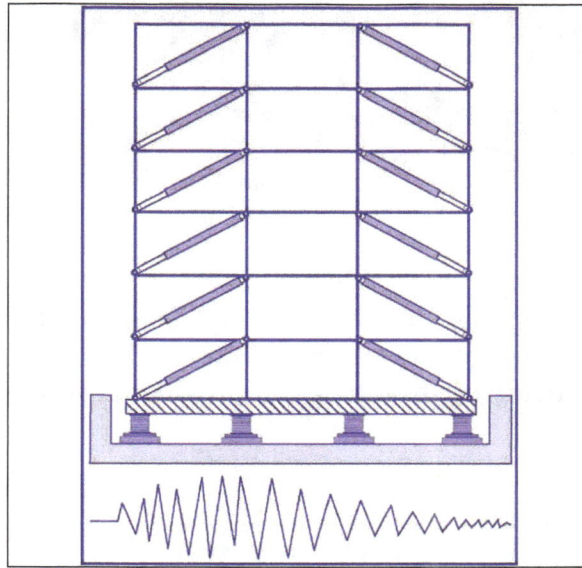

In order to properly understand how buildings and structures can stand up to earthquakes, extensive research has also been conducted on earthquakes.

In order to obtain an in depth knowledge concerning the initiation and behavior of earthquakes, it is essential to ascertain the mechanical properties and frictional characteristics of the crust of the earth. Observations from space have clarified the complete cycle of earthquake, including the silent accumulation of strain, transfer of stress between faults, release of strain, and failure of faults. Measurements on boundary zones of tectonic plates have explained the interaction of faults across hundreds of kilometers. Study of the stress transients that take place after earthquakes will determine the possibility of future earthquakes at other sites in the system. These studies have provided scientific explanations related to earthquake engineering and resulted in revision of concepts and practical application.

References

- Earthquake-geology, science: britannica.com, Retrieved 14 June, 2019

- Four-major-types-of-earthquakes: inventionsky.com, Retrieved 28 August, 2019

- Kato, Teruyaki (2011). "Slow earthquake". In Gupta, Harsh K. (ed.). Encyclopedia of Solid Earth Geophysics (2 ed.). Dordrecht: Springer. pp. 1374–1382. ISBN 978-90-481-8701-0. Retrieved 2013-04-07

- Rosakis, A.J.; Xia, K.; Lykotrafitis, G.; Kanamori, H. (2009). "Dynamic Shear Rupture in Frictional Interfaces: Speed, Directionality and Modes". In Kanamori H. & Schubert G. (ed.). Earthquake Seismology. Treatise on Geophysics. 4. Elsevier. pp. 11–20. doi:10.1016/B978-0-444-53802-4.00072-5. ISBN 9780444534637

- Horálek, Josef; Fischer, Tomáš; Einarsson, Páll; Jakobsdótir, Steinunn (2015). "Earthquake swarms". In Beer, Michael; Kougioumtzoglou, Ioannis; Patelli, Eduardo; Au, Siu-Kui (eds.). Encyclopedia of Earthquake Engineering. Berlin: Springer. pp. 871–885. doi:10.1007/978-3-642-35344-4. ISBN 978-3-642-35343-7

- Earthquake-engineering-goals-technology-and-research-methods, structural-engineering- 41861: brighthubengineering.com, Retrieved 06 February, 2019

- Xia, K.; Rosakis, A.J.; Kanamori, H. (2005). "Supershear and sub-Rayleigh to Supershear transition observed in laboratory earthquake experiments" (PDF). Experimental Techniques. Retrieved 28 April 2012

Subfields of Seismology

Engineering seismology, earthquake forecasting, archaeoseismology, paleoseismology, fault mechanics and seismic tomography are some of the fields studied under seismology. This chapter delves into these fields of seismology to provide an in-depth understanding of the subject.

ENGINEERING SEISMOLOGY

Engineering Seismology is the study of Seismology as related to Engineering. This involves understanding the source, the size and the mechanisms of earthquakes, how the ground motion propagates from the source to the site of engineering importance, the characteristics of ground motion at the site and how the ground motion is evaluated for engineering design. This subject is therefore related to the hazard of earthquakes. The seismic hazard at a site cannot be controlled. It can only be assessed. In the same context, Earthquake Engineering is the subject of analysis and design of structures to resist stresses caused by the earthquake ground motion. Resisting the stresses imply either resisting without failure or yielding to the stresses gracefully without collapse. This subject is related to the vulnerability of built structures to seismic ground motion. The vulnerability is controlled by design. The decision to control the vulnerability of a structure is based on the economics of the situation and on the judgement about the acceptable risk to the community.

Therefore, the assessment of seismic risk is based on the seismic hazard, the vulnerability and the value of the loss. This is expressed by the relation:

Risk = Hazard * Vulnerability * Value.

The value may be taken in the sense of cost of replacement.

In this context, "Seismic Hazard" is defined as the probability of occurrence of a ground motion of a given size within a given period of time at the site of interest. This will depend on the possible sources of earthquakes within a reasonable distance of the site and the seismic activity of these sources in relation to size and time.

The "Vulnerability" is a measure of the probability of damage (loss) to the structure to a ground motion of a given size. Different structures have different vulnerability curves.

Whereas the focus in the past was on what to do after a catastrophic earthquake, nowadays attention is shifting more and more to prevention, comprising the implementation of measures designed to mitigate risks. To this end, engineering seismology aims to lay down and develop corresponding foundations and methods used to enable the assessment of site-specific seismic hazards. It combines the disciplines of historical seismology, strong motion seismology and location effects, numerical modelling and earthquake-induced phenomena, and then crystallises that knowledge into basic practical guidelines for civil engineering, spatial planning and setting technical standards for construction.

Earthquake Catalogues and Seismic Observations

A robust earthquake catalogue provides a basis for making statistical statements about the probability of earthquakes occurring in a certain area. Historical details of the locations and strength of earthquakes are derived primarily from observations of their impact on nature, people and buildings. The consequences are described along with tremor's macroseismic intensity, the distribution of which serves to calculate the magnitude of historical earthquakes. Since having a comprehensive earthquake catalogue based on homogeneous magnitudes is an essential prerequisite for forecasting seismic hazards, gathering macroseismic intensity data is also an important task in the modern era of instrumental recording, because it helps us improve the calibration of historical quakes in the future.

Microzonation – Looking into the Local Geological Underground

Compared with solid, rocky ground, soft soils like valley floors, riverbanks and lake shores can amplify seismic tremors by a factor of up to 10 (in extreme cases). Consequently, in addition to producing seismic hazard maps showing regional differences in risk levels, the likelihood of site-specific underground tremors also has to be ascertained, in so-called microzonation studies. These studies entail the geological and geotechnical mapping of unconsolidated sediments, the analysis of slope stability and the potential for soil liquefaction, the use of geophysical measurements to predict amplifications of seismic waves, and numerical simulations calibrated to earthquake recordings. Several such studies have been carried out, concentrating especially on the regions around Basel, Lucerne, Sion and Visp in the canton of Valais, Bucharest in Romania and parts of the Egyptian capital, Cairo.

Risk Analysis – An Instrument for Planning Earthquake Mitigation

A medium earthquake risk combined with a dense population and high value concentration means that Switzerland faces a high seismic hazard. The quantification of the actual risk level is based on maximally realistic earthquake damage scenarios. Building vulnerability is also taken into account in so-called 'fragility curves' and combined with expectations about local seisms. Scenarios like this enable forecasts to be made about the potential physical and financial damage and numbers of fatalities, injuries and homeless people in the wake of an earthquake. Another key point entails the accurate factoring in of uncertainties when calculating damage scenarios. At the time of writing, all school buildings in the canton of Basel were being analysed using such a risk model, focussing in particular on issues associated with the cost-effectiveness of seismic retrofitting measures.

EARTHQUAKE FORECASTING

Earthquake forecasting is a branch of the science of seismology concerned with the probabilistic assessment of general earthquake seismic hazard, including the frequency and magnitude of damaging earthquakes in a given area over years or decades. While forecasting is usually considered to be a type of prediction, earthquake forecasting is often differentiated from earthquake prediction, whose goal is the specification of the time, location, and magnitude of future earthquakes with sufficient precision that a warning can be issued. Both forecasting and prediction of earthquakes are distinguished from earthquake warning systems, which upon detection of an earthquake, provide a real-time warning to regions that might be affected.

In the 1970s, scientists were optimistic that a practical method for predicting earthquakes would soon be found, but by the 1990s continuing failure led many to question whether it was even possible. Demonstrably successful predictions of large earthquakes have not occurred and the few claims of success are controversial. Consequently, many scientific and government resources have been used for probabilistic seismic hazard estimates rather than prediction of individual earthquakes. Such estimates are used to establish building codes, insurance rate structures, awareness and preparedness programs, and public policy related to seismic events. In addition to regional earthquake forecasts, such seismic hazard calculations can take factors such as local geological conditions into account. Anticipated ground motion can then be used to guide building design criteria.

Methods for Earthquake Forecasting

Methods for earthquake forecasting generally look for trends or patterns that lead to an earthquake. As these trends may be complex and involve many variables, advanced statistical techniques are often needed to understand them, therefore these are sometimes called statistical methods. These approaches tend to have relatively long time periods, making them useful for earthquake forecasting.

Elastic Rebound

Even the stiffest of rock is not perfectly rigid. Given a large force (such as between two immense tectonic plates moving past each other) the earth's crust will bend or deform. According to the elastic rebound theory of Reid, eventually the deformation (strain) becomes great enough that something breaks, usually at an existing fault. Slippage along the break (an earthquake) allows the rock on each side to rebound to a less deformed state. In the process energy is released in various forms, including seismic waves. The cycle of tectonic force being accumulated in elastic deformation and released in a sudden rebound is then repeated. As the displacement from a single earthquake ranges from less than a meter to around 10 meters (for an M 8 quake), the demonstrated existence of large strike-slip displacements of hundreds of miles shows the existence of a long running earthquake cycle.

Characteristic Earthquakes

The most studied earthquake faults (such as the Nankai megathrust, the Wasatch fault, and the San Andreas fault) appear to have distinct segments. The *characteristic earthquake* model postulates that earthquakes are generally constrained within these segments. As the lengths and other properties of the segments are fixed, earthquakes that rupture the entire fault should have similar characteristics. These include the maximum magnitude (which is limited by the length of the rupture), and the amount of accumulated strain needed to rupture the fault segment. Since continuous plate motions cause the strain to accumulate steadily, seismic activity on a given segment should be dominated by earthquakes of similar characteristics that recur at somewhat regular intervals. For a given fault segment, identifying these characteristic earthquakes and timing their recurrence rate (or conversely return period) should therefore inform us about the next rupture; this is the approach generally used in forecasting seismic hazard. Return periods are also used for forecasting other rare events, such as cyclones and floods, and assume that future frequency will be similar to observed frequency to date.

Extrapolation from the Parkfield earthquakes of 1857, 1881, 1901, 1922, 1934, and 1966 led to a forecast of an earthquake around 1988, or before 1993 at the latest (at the 95% confidence interval), based on the characteristic earthquake model. Instrumentation was put in place in hopes of detecting precursors of the anticipated earthquake. However, forecasted earthquake did not occur until 2004. The failure of the Parkfield prediction experiment has raised doubt as to the validity of the characteristic earthquake model itself.

Seismic Gaps

At the contact where two tectonic plates slip past each other every section must eventually slip, as (in the long-term) none get left behind. But they do not all slip at the same time; different sections will be at different stages in the cycle of strain (deformation) accumulation and sudden rebound. In the seismic gap model the "next big quake" should be expected not in the segments where recent seismicity has relieved the strain, but in the intervening gaps where the unrelieved strain is the greatest. This model has an intuitive appeal; it is used in long-term forecasting, and was the basis of a series of circum-Pacific (Pacific Rim) forecasts in 1979 and 1989–1991.

However, some underlying assumptions about seismic gaps are now known to be incorrect. A close examination suggests that "there may be no information in seismic gaps about the time of occurrence or the magnitude of the next large event in the region"; statistical tests of the circum-Pacific forecasts shows that the seismic gap model "did not forecast large earthquakes well". Another study concluded that a long quiet period did not increase earthquake potential.

ARCHAEOSEISMOLOGY

Archaeoseismology is the study of past earthquakes deriving from the analysis of archaeological sites. Such analyses reveal information about seismic events that have not been historically recorded. Such data can also help to document seismic risk in areas subject to extremely destructive earthquakes. In 1991, an international conference held in Athens marked the beginning of modern research in the field of archaeoseismology, described as a "study of ancient earthquakes, and their social, cultural, historical and natural effects".

Archeological Record

The archaeological record can carry three different types of evidence of seismic activity:

- The archaeological remains are displaced due to the movement of an active fault.

- The remains and artefacts contained in destruction deposits, associated with the decline of soil or seismic vibration, can be used the dating of earthquake damage. Other archaeological evidence, such as repairs, abandonment of an archaeological site or architectural changes, can help in identifying ancient earthquakes.

- Ancient buildings and other man-made structures can be studied for signs of ancient seismic disaster, often associated with soil vibration.

PALEOSEISMOLOGY

Paleoseismology looks at geologic sediments and rocks, for signs of ancient earthquakes. It is used to supplement seismic monitoring, for the calculation of seismic hazard. Paleoseismology is usually restricted to geologic regimes that have undergone continuous sediment creation for the last few thousand years, such as swamps, lakes, river beds and shorelines.

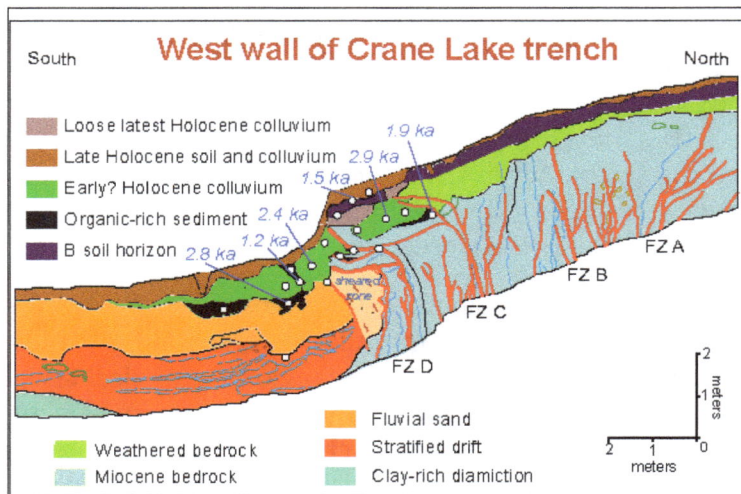

Sketch of trench wall.

Effects of tsunami caused by an earthquake.

In this typical example, a trench is dug in an active sedimentation regime. Evidence of thrust faulting can be seen in the walls of the trench. It becomes a matter of deducting the relative age of each fault, by cross-cutting patterns. The faults can be dated in absolute terms, if there is dateable carbon, or human artifacts.

Seismite formed by liquefaction of sediments during a Late Ordovician earthquake.

Many notable discoveries have been made using the techniques of paleoseismology. For example, there is a common misconception that having many smaller earthquakes can somehow 'relieve' a major fault such as the San Andreas Fault, and reduce the chance of a major earthquake. It is now known (using paleoseismology) that nearly all the movement of the fault takes place with extremely large earthquakes. All of these seismic events (with a moment magnitude of over 8), leave some sort of trace in the sedimentation record.

Another famous example involves the megathrust earthquakes of the Pacific Northwest. It was thought for some time that there was low seismic hazard in the region because relatively few modern earthquakes have been recorded. It was thought that the Cascadia subduction zone was merely sliding in a benign manner.

All of these comforting notions were shattered by paleoseismology studies showing evidence of extremely large earthquakes (the most recent being in 1700), along with historical tsunami records. In effect, the subduction zone under British Columbia, Washington, Oregon, and far northern California, is perfectly normal, being extremely hazardous in the long term, with the capability of generating coastal tsunamis of several hundred feet in height at the coast. These are caused by the interface between the subducted sea floor stressing the overlaying coastal soils in compression. Periodically a slip will occur which causes the coastal portion to reduce in elevation and thrust toward the west, leading to tsunamis in the central and eastern north Pacific Ocean (with several hours of warning) and a reflux of water toward the coastal shore, with little time for residents to escape.

FAULT MECHANICS

Fault mechanics is a field of study that investigates the behavior of geologic faults.

Behind every good earthquake is some weak rock. Whether the rock remains weak becomes an important point in determining the potential for bigger earthquakes.

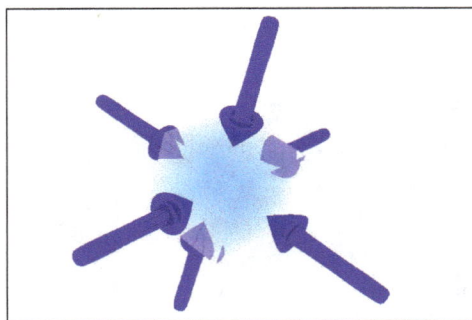

An element of rock under stress.

On a small scale, fractured rock behaves essentially the same throughout the world, in that the angle of friction is more or less uniform (see Fault friction). A small element of rock in a larger mass responds to stress changes in a well defined manner: if it is squeezed by differential stresses greater than its strength, it is capable of large deformations. A band of weak, fractured rock in a competent mass can deform to resemble a classic geologic fault. Using seismometers and earthquake location, the requisite pattern of micro-earthquakes can be observed.

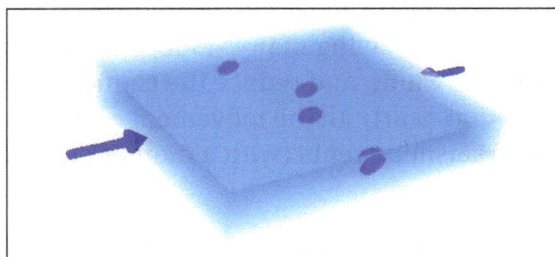

Penny-shape cracks in rock.

For earthquakes, it all starts with an embedded penny-shaped crack as first envisioned by Brune. As illustrated, an earthquake zone may start as a single crack, growing to form many individual cracks and collections of cracks along a fault. The key to fault growth is the concept of a "following force", as conveniently provided for interplate earthquakes, by the motion of tectonic plates. Under a following force, the seismic displacements eventually form a topographic feature, such as a mountain range.

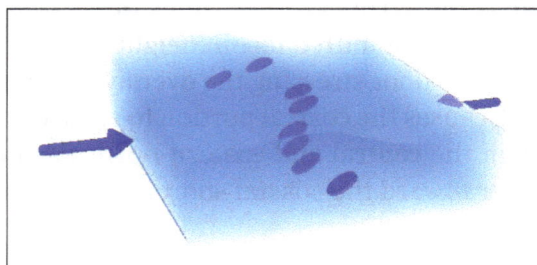

Following forces forming a mountain range.

Intraplate earthquakes do not have a following force, and are not associated with mountain building. Thus, there is the puzzling question of how long any interior active zone has to live. For, in a solid stressed plate, every seismic displacement acts to relieve (reduce) stress; the fault zone should come to equilibrium; and all seismic activity cease. One can see this type of arching "lockup" in many natural processes.

In fact, the seismic zone (such as the New Madrid Fault Zone) is ensured eternal life by the action of water. As shown, if we add the equivalent of a giant funnel to the crack, it becomes the beneficiary of stress corrosion (the progressive weakening of the crack edge by water). If there is a continuing supply of new water, the system does not come to equilibrium, but continues to grow, ever relieving stress from a larger and larger volume.

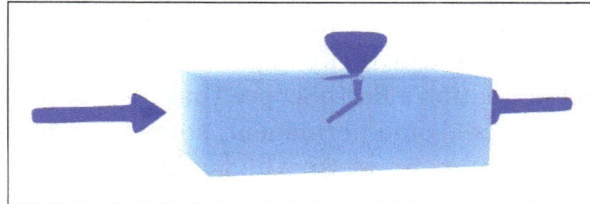

Fresh water continually being injected.

Thus the prerequisite for a continuing seismically active interior zone is the presence of water, the ability of the water to get down to the fault source (high permeability), and the usual high horizontal interior stresses of the rock mass. All small earthquake zones have the potential to grow to resemble New Madrid or Charlevoix.

SEISMIC TOMOGRAPHY

Seismic tomography is a method for reconstruction of continuous distribution of seismic parameters in 1D, 2D, 3D, or 4D (space and time) using the characteristics of seismic waves traveling between sources and receivers. Seismic parameters to be found in tomographic inversion are in most cases velocities of P and S seismic waves (P and S velocities). For volcanoes, one of the key parameters appears to be the Vp/Vs ratio which can be used to evaluate the content of fluids and melts. Besides the velocity distributions, seismic tomography may provide the information on the anisotropy of seismic parameters which helps studying regional stresses and space-oriented geological structures.

In recent years, a different sort of imaging technique has done the same for geophysicists. Seismic tomography allows them to detect and depict subterranean features.

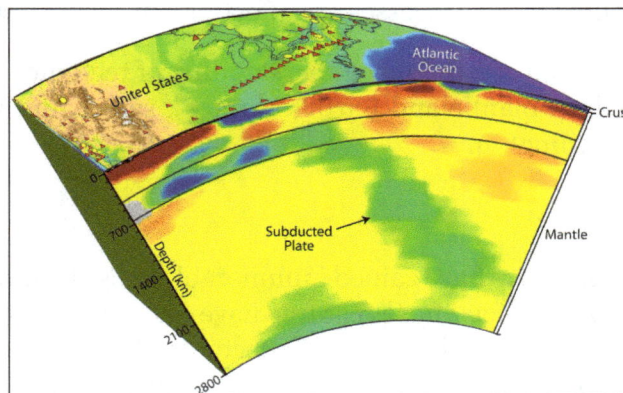

Data gathered by a network of seismic instruments (red) have enabled researchers to discern a region of relatively cold, stiff rock (shades of green and blue) beneath eastern North America.
This is likely to be the remnants of an ancient tectonic plate.

The advent of the approach has proven to be a boon for researchers looking to better understand what's going on beneath our feet. Results have offered myriad insights into environmental conditions within the Earth, sometimes hundreds or even thousands of kilometers below the surface. And in some cases, the technique offers evidence that bolsters models of geophysical processes long suspected but previously only theorized, researchers say.

Seismic tomography "lets us image Earth's structures at all sorts of scales," says Jeffrey Freymueller, a geophysicist at Michigan State University in East Lansing and director of the national office of the National Science Foundation's EarthScope. That 15-year program, among other things, operates an array of seismometers—some permanent, some temporary—that has collected data across North America. Among its more impressive finds: the remnants of an ancient tectonic plate sitting deep below North America and a plume of buoyant material fueling a well-known geothermal hot spot.

Dissecting the Earth

Tomography, roughly translated from Greek, means "writing by slices". Researchers relish this ability to take digital models of three-dimensional (3D) objects and slice through them to create cross-sections—to virtually dissect them from any angle. Both medical tomography and seismic tomography use large arrays of sensors to collect energy that has traveled through a given body. Medical tomography typically uses differences in the amounts of transmitted energy to create images with blacks, whites, and shades of gray.

But seismic tomography uses differences in the speed of seismic waves as they travel through Earth to construct its 3D model. In general, vibrations travel more slowly through rocks that are hotter or less dense, contain hydrated minerals, or are partially melted. On the other hand, seismic waves travel more quickly through rocks that are colder, denser, and drier. By knowing the precise time at which a distant earthquake occurred, as well as the times at which vibrations from that temblor arrived at each seismometer in a network, researchers can "invert" the data and map out the portions of the planet that those seismic waves had traveled through.

Before seismic tomography came along, geophysicists could only imagine what might be happening deep within Earth. The ability to probe thousands of kilometers underground can help researchers better decipher how those processes are affecting our planet's surface.

"Seismic tomography has revolutionized our understanding of tectonics and allows us to identify connections between the deep mantle and Earth's surface", says Laura Webb, a geologist at the University of Vermont in Burlington.

Long Time Buried

Using EarthScope data, researchers have gained innumerable insights into what lies beneath North America—and the geophysical effects those features have had, and are still having, on the continent. Many of these result from the subduction of a tectonic plate that began off North America's Pacific coast more than 165 million years ago. Although seismic tomography shows that the western edge of that ancient slab of ocean crust—which geologists have dubbed the Farallon Plate—still lies offshore, the bulk of it lies beneath the western United States. Fluids that were squeezed from the slab as it was shoved eastward beneath the continent rose to hydrate the underside of Earth's crust. Later, the languid motion of the underlying mantle buoyed the crust upward, gradually elevating the Colorado Plateau. The results of that process are indeed impressive, Webb notes, "That steady boost in the crust, over time, enabled the region's rivers to carve spectacular canyons".

Farther east, remnants of the Farallon Plate sit beneath the Midwest, where they've shed even more water to create a weak zone that stretches from just below the crust there to depths of around 200 kilometers, says Webb. That weak spot, not coincidentally, lies right beneath the New Madrid Fault Zone—which spawned some of America's largest earthquakes about two centuries ago. As the ancient slab slowly sinks, mantle flow around it creates a downward suction that stresses and deforms the overlying crust. Those stresses, if large enough, can trigger earthquakes. "Seismic tomography has shed new light on how this region can be so seismically active despite being far from tectonic plate boundaries where most large earthquakes occur", Webb notes.

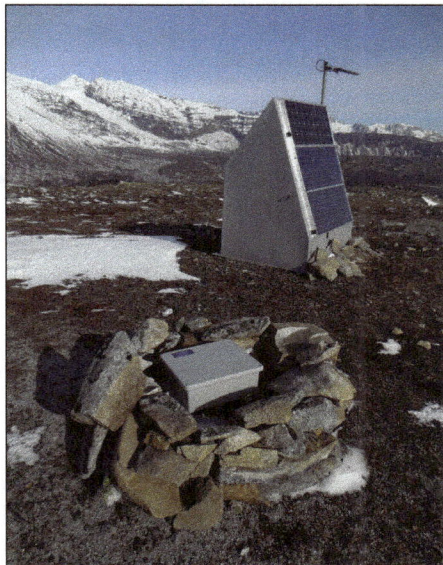

A continent-wide network of seismometers like this one, which was installed in south-central Alaska in 2016, helps researchers probe Earth's inner structure. Image credit: EarthScope National Office, a National Science Foundation funded project/Max Kaufman. EarthScope scientists study the structure and evolution of the North American continent using three primary observatories, the Plate Boundary Observatory, US Array, and the San Andreas Observatory at Depth.

From Titanic to Tiny

But seismic tomography can discern things far smaller than kilometers-thick slabs of subducted material. A slowdown in the seismic waves passing beneath Yellowstone National Park provides evidence for a deep-rooted plume of warm, buoyant material rising to the surface there. Geophysicists have long theorized that such plumes fuel volcanic activity at so-called hot spots around the world. Similarly, seismic tomography has offered views into the mid-to-low-level portions of Earth's crust beneath the park. There, a substantial slowing of seismic waves betrays the presence of a 46,000-cubic-kilometer blob of partially melted rock that connects the deep mantle plume to the shallow magma reservoir that's the heat source for the region's famed geysers.

On an even smaller scale and using a small network of a few dozen seismometers, researchers mapped out parts of the plumbing system beneath Mount Erebus, a volcano in Antarctica. During the 2008–2009 field season, the team set off a dozen small blasts on or near the peak, which was surrounded by a roughly 4-kilometer-by-4-kilometer network of 23 seismic instruments that had been deployed the previous summer. The seismic data from those blasts revealed that a large blob of magma—which in some places slowed down seismic waves by as much as 1 kilometer per second—lies beneath the northwestern slope of the volcano. The tubes that occasionally channel molten rock to the surface during eruptions are too small to be discerned by the analyses, the scientists report, but the volcano's magma chamber, which lies at least 500 meters below the surface, shows up clearly. Long-term studies of active volcanoes could reveal how changes in the size and shape of those peaks' magma reservoirs correlate with eruptions.

One of seismic tomography's most impressive coups, however, may be spotting regions of rock kilometers below Earth's surface where the minerals' atoms may be arranged differently from those in surrounding regions or at different depths. Using data gathered by researchers in previous studies, Fan-Chi Lin, a geophysicist at the University of Utah in Salt Lake City, and his colleagues used tomography to image Earth's crust beneath the Yellowstone caldera in Wyoming and the Long Valley caldera in California.

For their study, the researchers added an extra layer to the tomographic analysis: They not only estimated differences in the overall velocities of seismic waves passing through the crust, but they also looked at the differences between the speeds of horizontally polarized seismic waves compared with those that were vertically polarized.

Those findings suggest that the rocks lying beneath the Yellowstone and Long Valley calderas at depths of between 5 and 18 kilometers were likely arranged in a number of horizontal layers. Some of those layers could be only a few meters thick, says Lin, and as much as 6% of the rock within them could be melted. And that slushiness, in turn, could have big implications for how quickly, on geological timescales, those reservoirs of mushy rock could mobilize to generate future eruptions, the researchers say. If geophysicists based their long-term predictions on analyses that didn't include layers of partially molten rock, they could dramatically underestimate the time it could take for that rock to move into a magma reservoir that could someday generate a major eruption.

Just beneath the Surface

Researchers don't need to use seismic waves from distant earthquakes—or from blasts they triggered themselves—to perform seismic tomography. They can also use the near-continual low-level shimmies that shake the ground and sometimes interfere with measurements of distant quakes. Geophysicists can use this "seismic hum" the same way that photographers use dim light at night: Added up over time, the energy is sufficient to generate an image, explains Keith Koper, another geophysicist at the University of Utah in Salt Lake City. His analyses have shown that the seismic vibrations generated by waves on six lakes in North America and in China—even lakes that cover less than 300 square kilometers—can shake the ground up to 30 kilometers away from those bodies of water.

Ocean surf, including the waves pounding the Pacific coastline, also generates seismic hum. Researchers in southern California recently used data recorded in 2015 by 315 seismometers across the region to map out fault zones and other geological features in the area. Many of those features—such as the deep structure of the San Andreas, Hayward, Whittier, and other fault zones—were known from previous tomographic studies. But by including new information, such as the size of the seismic waves as well as their arrival times, the new tomographic image has a much better resolution of smaller features that lie within the upper 3 kilometers of the crust. This enhanced mapping of fault zones and sediment depths could improve future analyses of the region's seismic risk, the team suggests.

Following up on earlier field work, Koper and his team deployed a few seismometers on the floor of Yellowstone Lake last summer. Those instruments, as well as others in a shore-based network, will record vibrations generated by the lake's waves as well as those generated by innumerable small quakes in the Yellowstone area. Using that data, Koper and his colleagues will try to map the plumbing system that feeds hydrothermal vents beneath the northern end of Yellowstone Lake.

"Those vents are, in essence, underwater versions of the geysers seen elsewhere in Yellowstone", Koper notes. Such efforts are just another example of how discerning what's hidden beneath Earth's surface can help researchers better understand the processes unfolding at ground level.

The National Science Foundation's funding for EarthScope has ended, but the program's legacy will last far into the future. About 80 of the nearly 300 seismometers temporarily installed in Alaska over the past few years will be "adopted" by the US Geological Survey or the University of Alaska Fairbanks, says Robert Woodward, director of instrumentation services for Incorporated Research Institutions for Seismology (IRIS), a Washington, DC-based consortium of universities. The remainder will be decommissioned and removed over the next couple of field seasons. IRIS will recalibrate and loan out those instruments to seismologists pursuing short- and long-term projects in the coming years, says Woodward.

PERMISSIONS

We would like to thank the editorial team for lending their expertise to make the book truly unique. They have played a crucial role in the development of this book. Without their invaluable contributions this book wouldn't have been possible. They have made vital efforts to compile up to date information on the varied aspects of this subject to make this book a valuable addition to the collection of many professionals and students.

This book was conceptualized with the vision of imparting up-to-date and integrated information in this field. To ensure the same, a matchless editorial board was set up. Every individual on the board went through rigorous rounds of assessment to prove their worth. After which they invested a large part of their time researching and compiling the most relevant data for our readers.

The editorial board has been involved in producing this book since its inception. They have spent rigorous hours researching and exploring the diverse topics which have resulted in the successful publishing of this book. They have passed on their knowledge of decades through this book. To expedite this challenging task, the publisher supported the team at every step. A small team of assistant editors was also appointed to further simplify the editing procedure and attain best results for the readers.

Apart from the editorial board, the designing team has also invested a significant amount of their time in understanding the subject and creating the most relevant covers. They scrutinized every image to scout for the most suitable representation of the subject and create an appropriate cover for the book.

The publishing team has been an ardent support to the editorial, designing and production team. Their endless efforts to recruit the best for this project, has resulted in the accomplishment of this book. They are a veteran in the field of academics and their pool of knowledge is as vast as their experience in printing. Their expertise and guidance has proved useful at every step. Their uncompromising quality standards have made this book an exceptional effort. Their encouragement from time to time has been an inspiration for everyone.

The publisher and the editorial board hope that this book will prove to be a valuable piece of knowledge for students, practitioners and scholars across the globe.